Introducing Metamorphism

Other Titles in this Series:

Introducing Astronomy
Introducing Geology – A Guide to the World of Rocks
(Third Edition)
Introducing Geomorphology
Introducing Meteorology ~ A Guide to the Weather
Introducing Mineralogy
Introducing Oceanography
Introducing Palaeontology ~ A Guide to Ancient Life
Introducing Sea Level Change (forthcoming)
Introducing Sedimentology
Introducing Stratigraphy (forthcoming)
Introducing Tectonics, Rock Structures and Mountain Belts
Introducing the Planets and their Moons
Introducing Volcanology ~ A Guide to Hot Rocks

For further details of these and other Dunedin Earth and Environmental Sciences titles see
www.dunedinacademicpress.co.uk

INTRODUCING
METAMORPHISM

Ian Sanders

DUNEDIN

EDINBURGH ◆ LONDON

Published by
Dunedin Academic Press Ltd
Hudson House
8 Albany Street
Edinburgh EH1 3QB
Scotland

www.dunedinacademicpress.co.uk

ISBNs
9781780460642 (Paperback)
9781780465982 (ePub)
9781780465999 (Kindle)

British Library Cataloguing in Publication data
A catalogue record for this book is available from the British Library

Typeset by Makar Publishing Production, Edinburgh, Scotland
Printed in Poland by Hussar Books

Contents

vii

Preface

Students of geology do not normally start with metamorphic rocks. They normally learn about these rocks later in their course, after having become familiar with the internal structure of the planet, the theory of plate tectonics, the chemistry of minerals, and the skill of using the polarizing microscope. Introducing metamorphism to a readership who may have little prior knowledge of these topics presents challenges.

In an attempt to meet the challenges, a selection of relevant background information is provided as appendices. Appendix 1 deals with the Earth's interior and plate tectonics. Appendix 2 reviews minerals. Appendix 3 outlines the use of the polarizing microscope. The content of each appendix has been carefully chosen with metamorphism in mind, and is cross-referenced to the main text. Readers who are new to geology might like to look through these appendices before beginning chapter 1

This book is not a comprehensive account of metamorphism, but is biased towards the author's interests and experience. The aim is to give an overall sense of what metamorphism entails through describing a broad selection of metamorphic rocks, through explaining the methods (particularly polarized light microscopy) used to investigate them, and particularly through addressing all kinds of questions about how the rocks came to be the way they are found today.

Following a general introduction (chapter 1), the text is organized into four chapters. The first of these, chapter 2, is easily the longest, and is largely descriptive. The other three focus on interpretation. Chapter 3 asks about the processes that lie behind features seen on the scale of a hand specimen or smaller, chapter 4 extends the questioning to processes that relate to the geological setting and timing of metamorphism, and chapter 5 aims to quantify the pressure and temperature conditions during metamorphism in two specific case studies which the author knows well, and continues to find fascinating.

All terms in **bold font** in the appendices and in the main text are defined in a Glossary at the end of the book.

Acknowledgements

x

I am most grateful to Anthony Kinahan of Dunedin Press for inviting me to add *Metamorphism* to Dunedin's growing stable of 'introductions' to the Earth and environmental sciences, and for his patience, support, and good humour during the preparation of the book. Elaine Cullen kindly drafted the diagrams, and her suggestions for improving them were always appropriate and welcome. Science Foundation Ireland funded the Irish University Rock collection of Virtual Microscope images that feature in the book, and which were prepared by Andy Tindle at the Open University. The examples I have chosen and the explanations for their origins came to me over many years through dialogue and shared observations with my teachers, colleagues, students and friends. I am grateful to them all, and particularly to those who generously let me use their photographs. I am also grateful to Ralf Halama and an anonymous reviewer, whose insights and advice led to significant improvements in the text. The book would never have been completed without the sustained encouragement, understanding and cajoling of my wife Sheila. Her support has been immeasurable.

Ian Sanders
Trinity College Dublin
March 2018

1 Introduction

1.1 What is metamorphism?

Metamorphism (from the Greek words *meta* = change and *morphos* = form) is a geological process that changes pre-existing **igneous** and **sedimentary rocks** into new rocks – metamorphic rocks – that look quite different from the rocks they started out as. Metamorphism can even change pre-existing metamorphic rocks into new, and different looking, metamorphic rocks. When rocks are changed like this they are said to become metamorphosed, but one should note that the related word *metamorphosis* has no place in geology. It applies to other kinds of change, such as from a tadpole to a frog, and is not a synonym for metamorphism.

Metamorphism takes place underground, usually deep underground at a high temperature. Since it cannot be watched as it happens, it is often portrayed as being more difficult to grasp than the formation of igneous or sedimentary rocks. Igneous rocks can be seen being made when, for example, volcanic **lava** cools and hardens. The formation of sedimentary rocks can be watched as grains of sand, for example, are moved by rivers and deposited on the seabed. Yet the very fact that metamorphic rocks originate 'out of sight' makes them all the more intriguing. The detective work involved in figuring out how they were made can be both challenging and rewarding.

1.1.1 Protoliths

An igneous or sedimentary rock that becomes metamorphosed is called a **protolith** (from the Greek words *proto* = first and *lithos* = stone). While there are many kinds of protolith, only six are important for the purpose of introducing metamorphism. These are the three common kinds of sedimentary rock, called **sandstone**, **shale** and **limestone**, the two common kinds of igneous rock, **basalt** and **granite**, and the important but less well-known rock called **peridotite** (Fig. 1.1).

The first five protoliths are common in the Earth's **crust**, whereas the sixth one, peridotite, is the main kind of rock in the Earth's **mantle**. Readers who would like to remind themselves

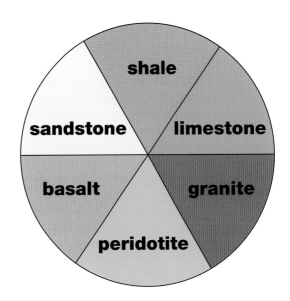

Figure 1.1 The six main protoliths.

of the nature of the Earth's crust and mantle should turn now to section A1.1 of Appendix 1. Peridotite is important because it is the primary rock from which the other five protoliths have been derived, over time, by geological processes that operate at the surface and beneath it. These processes are touched on in this chapter, and are explained in more detail in Appendix 1 and chapter 2. At the end of chapter 2 they are summarized in a flow chart called the **rock cycle**.

When a protolith becomes metamorphosed, its appearance generally changes in two distinct ways. Firstly, it usually develops new **minerals**, and secondly, it always develops a new **texture**.

1.1.2 Changes to the minerals

Minerals are a rock's ingredients. They are the solid chemical compounds from which the grains (individual particles) of a rock are made. Examples of minerals include quartz, garnet

2

and mica. Each has its own distinctive features and its own chemical composition. Most metamorphic rocks are aggregates of two, three, four or more different minerals. The list of minerals in a rock is known as the rock's **mineral assemblage**.

The minerals in a rock change because, during metamorphism, the pressure and temperature beneath the surface change. The existing minerals become unstable together and react chemically with each other to produce new minerals that are stable together under the new pressure and temperature conditions. The meaning of mineral stability is explained in chapter 3.

The number of different minerals that can occur in metamorphic rocks is enormous, but only about two dozen of them account for most of the rocks described in this book, with a further ten, called **accessory minerals**, being widespread but never abundant. These minerals are introduced one at a time in chapter 2, as their names arise. An emphasis is placed on how they can be recognized and on which chemical elements each contains.

Just ten chemical elements are needed to make the main two-dozen minerals, namely hydrogen (symbol H), carbon (C), oxygen (O), sodium (Na), magnesium (Mg), aluminium (Al), silicon (Si), potassium (K), calcium (Ca), and iron (Fe). The way in which these elements team up to make minerals is explained from first principles in Appendix 2.

It is helpful to know the elements in each mineral because then, if the minerals in a rock can be recognized, the overall chemistry of the rock can be estimated. This allows the protolith to be worked out, because each kind of protolith has its own distinctive chemical signature, which generally does not change during metamorphism.

However, there is an important exception to this constancy of chemical composition. This is a change in the amounts of water (H_2O) and/or carbon dioxide (CO_2) that are chemically bound in the rock's minerals. Minerals that contain chemically bound water are described as hydrous, and their chemical formulae contain the so-called hydroxyl group, (OH). Minerals that contain chemically bound carbon dioxide are known as carbonate minerals, and their chemical formulae have the carbonate group, (CO_3). Both water and carbon dioxide are examples of **fluids** (the term *fluid* covers liquids *and* gases), and they can be added to rocks, and can be lost from rocks, during metamorphism. Many examples of such gains and losses will be encountered in the book.

Perhaps the best evidence that fluid really was present during metamorphism is that it can be seen today, sealed up inside tiny cavities within mineral grains in some metamorphic rocks. Such fluid-filled cavities are known as **fluid inclusions** (Fig. 1.2). They are thought to have formed when small quantities of fluid became trapped inside mineral grains as they grew larger during metamorphism.

Since fluids can carry dissolved salts, then the introduction of fluids to rocks, and their loss from rocks, during metamorphism will cause any dissolved elements to move into, and out of, the rocks. If the resulting chemical changes to the rocks are substantial, the process is called **metasomatism** (pronounced meta-**soh**-ma-tism; from the Greek words *meta* = change and *soma* = body).

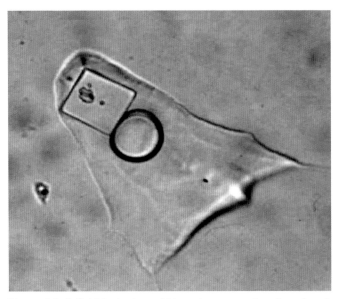

Figure 1.2 A fluid inclusion within a transparent wafer of rock seen through a microscope. The image is about 10 microns (10 thousandths of a millimetre) wide. The fluid is water. It is trapped in a cavity with a roughly triangular outline inside a grain of metamorphic quartz from an emerald deposit in Columbia. The water was very hot when it became trapped, and has since cooled and contracted, causing it to separate into liquid water and a bubble of water vapour. Also present in the water is a cube-shaped crystal of sodium chloride (square outline) that was precipitated during cooling. It shows that the liquid is not pure water but salty water. *Photo courtesy of Bruce Yardley.*

1.1.3 Changes to the texture

The second kind of change a rock undergoes during metamorphism is a change in its texture. The term texture needs a bit of explanation. To most people texture is the way cloth or other fabric feels when handled – whether, for example, it is silky, supple or coarse to touch. When applied to rocks, texture has a different and very specific meaning. It refers to the sizes and shapes of the mineral grains, and to the way the grains are orientated and distributed within the rock. A great variety of textures may be seen among metamorphic rocks, and these will be described and interpreted in chapter 2 and elsewhere in the book. For instance, grains may be too tiny to be seen with the naked eye or large enough to be spotted easily; grains with an elongate shape may be aligned parallel to each other or they may be randomly orientated; grains of a particular mineral may be evenly distributed throughout a rock, like dispersed raisins in a well-mixed fruit cake, or they may be concentrated in discrete clusters or layers.

During metamorphism the texture automatically changes when a new combination of minerals replaces an old one, but it can also change when existing mineral grains grow in size, and change their shapes, through a process called **grain growth** or **recrystallization**. With grain growth, small grains merge to become larger ones, for reasons that will be explained in chapter 3.

On the basis of their texture, rocks fall into two broad groups, known as **foliated** rocks and non-foliated rocks. In foliated rocks, grains of minerals with elongated shapes are aligned, and the rocks may also be layered in the same direction as the mineral alignment. Non-foliated rocks have no preferred orientation of elongated grains, and are not layered. Whether or not a rock is foliated depends on a factor called directed stress, which will be introduced below, in section 1.2.

1.1.4 Naming metamorphic rocks

Metamorphic rocks have a somewhat confusing choice of names. In fact, an individual rock may be given two or more different names that are equally correct. The choice arises because a rock's name can be based on its protolith, or its minerals, or its texture, or on some combination of these three.

Names based on the protolith are, perhaps, the simplest. Any metamorphic rock can be named by adding the prefix **meta** to the name of the protolith giving, for example, meta-limestone, metashale, and metabasalt. Most metamorphosed sedimentary rocks, however, already have their own widely used names relating to the protolith: **marble** is nearly always used instead of metalimestone, **metapelite** (pronounced meta-**pee**-lite) is often used as an alternative to metashale, and **metapsammite** (the 'p' is silent) is a popular alternative to metasandstone.

Names based on the rock's texture include several that relate to a foliated texture. These are **slate** or **phyllite** (pronounced **fill**-ite) if the foliated rock is very fine-grained, **schist** if it is medium or coarse-grained, and **gneiss** (pronounced nice, the 'g' is silent) if it is both coarse-grained and layered. In many cases these four kinds of foliated rock are metapelites.

Names based on a rock's minerals can be created by hyphenating one or more of the prominent minerals in the rock to a name like schist, or gneiss, or simply 'rock', giving, for example, garnet-mica-schist, quartz-feldspar-gneiss, or quartz-garnet-rock. A few rocks have special names that relate to their main minerals, like quartzite (made largely from quartz) or amphibolite (made mostly from a mineral called amphibole).

Examples of rocks with all the above names, and with additional names, will be described in chapter 2.

1.2 Metamorphic rocks – made under mountains

Since metamorphic rocks are formed beneath the Earth's surface, they are seen today only because, after they were made, the land became elevated and then eroded away to expose them. The great majority of metamorphic rocks fall under the heading of **regional metamorphic rocks** because they occur over extensive regions. These rocks were made, and are still being made, deep within the continental crust in the roots of major mountain ranges (Fig. 1.3).

A minority of metamorphic rocks are not from mountain roots, but were made in local settings beneath the ground, such as close to bodies of hot, shallow magma, or within deep fault zones, or below the impact sites of giant meteorites. Metamorphism in these local settings is introduced separately, after this section on metamorphism under mountains.

1.2.1 Mountain building

Mountain building is a protracted process; its formal name is **orogenesis**, or simply **orogeny** (from the Greek word *oros* = mountain). Orogeny is commonly preceded by the accumulation of many layers of sediment (mud, sand, and calcium

3

4

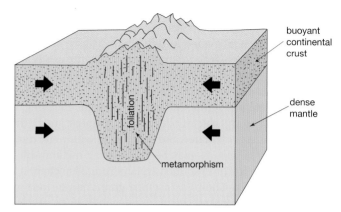

Figure 1.4 Simplified block diagram showing continental crust and underlying mantle in a place where orogenesis (mountain-building) is in progress. The continental crust thickened by directed stress (compressive forces shown by the black arrows) is buoyant. It 'floats' in the mantle, holding the mountains up.

Figure 1.3 Mountains in the Mont Blanc area of the Western Alps. Metamorphism is happening today in the hot, deep roots of this and all other major mountain ranges.

carbonate shells) in a place where the Earth's surface continues to subside over a long period of geological time. The weight of the upper layers of sediment exerts a huge downward pressure on the layers below. Also, the temperature gets higher with increasing depth because heat is continually produced by traces of **radioactive** uranium, thorium and potassium that are present in all rocks. However, the heat and pressure at this stage of burial will generally not be sufficient to cause metamorphism, but only enough to convert the loose accumulated sediments into hard sedimentary rocks. Mud becomes shale or mudstone, sand becomes sandstone, and shells become limestone, in a process known as **lithification** (from the Greek word *lithos* = stone). An additional geological factor is normally needed to induce metamorphism. This factor is the onset of a strong compressive force that arises when the local tectonic plates change their relative motions and begin to converge. The nature of **tectonic plates** and the theory behind their movement is outlined in section A1.2 of Appendix 1.

The compressive force is called **directed stress**, and it acts like a giant vice; it crumples and buckles the layers of

sedimentary rock (and also any igneous rocks that might have been injected into them as magma), squeezing them simultaneously upwards to form a mountain range and downwards, deeper into the mantle, to form the mountain's root (Fig. 1.4). It also squeezes the pre-existing rocks of the continental crust beneath the sediments. The entire continental crust might be doubled in thickness during this process.

As noted in Appendix 1, continental crust is less dense than the mantle, so it is buoyant. The mountain's root behaves like the underwater portion of an iceberg. Just as an iceberg floats in the sea with its top towering above the waterline, so the thickened continental crust floats, as it were, in the mantle with its top sticking up as mountains. This analogy must not be taken too far, though. While the mountains and their roots are imagined as floating in the mantle, the mantle is not liquid; it is not composed of molten rock, as some school textbooks suggest when inadvertently confusing *mantle* with *magma*. The mantle is known to be solid because, as is explained in section A1.1 of Appendix 1, seismic S-waves can travel through it.

1.2.2 Directed stress, pressure and temperature in a mountain's roots

As the rocks in the mountain's deep root become hot and squeezed, so they become metamorphosed. As the new minerals grow, the directed stress that causes the crust to become much thicker also causes the newly formed metamorphic rocks to become foliated. Slate, phyllite, schist and gneiss commonly develop. Images of these rocks are shown in chapter 2. The plane of the foliation is represented by vertical dashes in Figure 1.4, and is usually roughly at right angles to the direction in which the mountain's root is being squeezed.

Directed stress should not be confused with a related term, **lithostatic stress**. Lithostatic stress is the same thing as **pressure** (abbreviated 'P'). Pressure increases with depth and has the effect of making a rock denser by squashing it equally in all directions into a smaller volume. Thus, high pressure favours mineral combinations that are denser than those formed at low pressure. Pressure in metamorphism is usually measured in **kilobars** (kbar); a depth of 35km corresponds roughly to a pressure of 10kbar. Directed stress, in contrast, does not affect a rock's volume. Instead it causes a rock to change in shape by being flattened or sheared, as captured in the cartoon of stressed and pressurized divers in Fig. 1.5. The technical word for a change in shape is **strain**; so directed stress causes strain.

The temperature (abbreviated 'T') of metamorphism is denoted in a loose way by the term **grade**; metamorphism can be low-, medium-, or high-grade. Most geologists use the term

low-grade for temperatures between about 300°C and 500°C, medium-grade for temperatures between about 500°C and 700°C, and high-grade for temperatures between about 700°C and 900°C, as shown by the coloured areas in Figure 1.6. Slate and phyllite are low-grade rocks, schist is typically a medium-grade rock, and gneiss is usually a high-grade rock.

Describing rocks as low-, medium- or high-grade is not the only way of referring to the temperature conditions of their formation. Two other ways are by stating the so-called metamorphic **zone**, and by stating the so-called metamorphic **facies** (pronounced either fash-eez or fay-sheez). Metamorphic zones are based on key indicator minerals, like garnet, that develop in metapelite (metashale) as the grade increases. Metamorphic facies are based on distinctive combinations of minerals that develop in rocks of basaltic composition as the grade increases, and also as the pressure increases. Zones and facies will both be introduced fully in chapter 2. For the present, the names of three important facies – greenschist, amphibolite (pronounced am-**fib**-a-lite) and granulite – are shown in Figure 1.6. They correspond respectively to low-, medium- and high-grade conditions in rocks formed at intermediate pressures. Rocks formed at high pressures are different, and have their own facies names, as will be explained in chapters 2 and 4. Ways in which the P and T conditions of metamorphism can be quantified are considered in chapters 3 and 4, and especially in chapter 5.

Metamorphism has lower and upper temperature limits. At temperatures below about 300°C metamorphism barely gets started, and any changes tend to be imperceptible to the unaided eye. In sedimentary rocks the process of change at these temperatures is called **diagenesis** (pronounced dye-a-**gen**esis) rather than metamorphism. Diagenesis accompanies and follows lithification. Between roughly 200°C and 300°C it is sometimes called very-low-grade metamorphism, and is also called **anchimetamorphism** (pronounced **ank**imetamorphism). This temperature range is referred to as the **anchizone** (Fig. 1.6).

At the upper end of the temperature scale, high-grade metamorphism causes some rocks, particularly metapelite, to partially melt, and a kind of gneiss called **migmatite** is produced. The nature of migmatite is described in chapter 2, and its origin is explained in chapter 3.

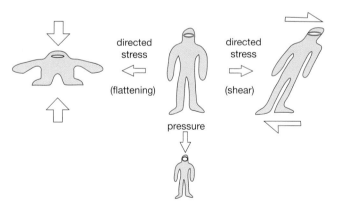

Figure 1.5 Cartoon highlighting the distinction between pressure (sometimes called lithostatic stress) and directed stress. The former causes a reduction in volume, the latter causes strain (change in shape), by flattening or shearing, with no change in volume.

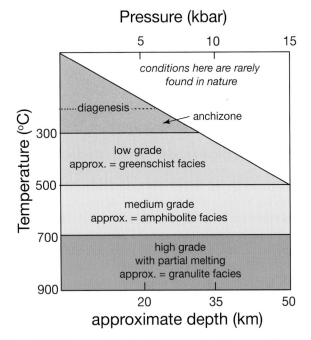

Figure 1.6 Diagram showing the approximate range of pressure (P) and temperature (T) conditions at which metamorphism occurs in the continental crust. Temperature bands are shown for low-grade, medium-grade, and high-grade metamorphism. P–T conditions in the white triangle are rare to absent in nature because temperature almost always increases with depth at more than about 10°C per km (equivalent to 500°C at 50km). For an explanation of other terms, see the text.

1.2.3 *Exhumation of a mountain's roots*

Mountain building is over when, after perhaps a few million years, the convergent **plate** motions cease. Erosion will then steadily reduce the height of the mountains, and the debris will be carried downhill to accumulate as sediment in neighbouring low-lying places. As the mountains are stripped away, their buoyant roots will continually adjust to the reduced load by floating upwards and maintaining the mountains, though at a lower height than before. Erosion will come to an end only when the root has entirely gone, and the continental crust has been restored to its original stable thickness, typically of about 35km. A huge volume of rock, as indicated in Figure 1.7, will by then have been removed, and the topography will have been reduced almost to sea level. An enormous swathe

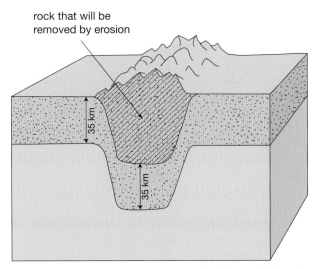

Figure 1.7 The same block diagram as in Figure 1.4, but here showing the volume of rock (diagonal lines) that will be removed by erosion once the lateral compression is over. After erosion is complete, metamorphic rocks from a considerable depth will be at the surface.

of metamorphic rocks that were formerly deep within the root will have become exposed at the surface. This swathe of rocks is known as a regional **metamorphic belt** or an **orogenic belt**. Rocks that were formed at higher grades, like schist and gneiss, will tend to be in the centre of the belt, with lower-grade rocks, like slate and phyllite, located towards the margins.

Metamorphic belts are typically tens to one or two hundred kilometres wide and several hundred, up to one or two thousand, kilometres in length. Metamorphic belts of various ages can be examined in different parts of the world today. One of them, called the Grampian belt, provides many of the examples of metamorphic rock described here. The Grampian belt extends diagonally in a NE to SW direction across Scotland and north-western Ireland. It is the vestige of a mountain range that existed about 470 million years ago.

1.3 Metamorphism in local settings

Contact, hydrothermal, dynamic and shock metamorphism are four kinds of metamorphism that, in contrast to regional metamorphism, are quite restricted in their geographical extent. They are briefly introduced here, and aspects of them are developed in later parts of the book.

1.3.1 Contact metamorphism

Contact metamorphism happens next to bodies of hot igneous rock beneath the ground. The heat from a body of molten granite 10km in diameter will, for example, affect the surrounding cool rock (called the **country rock**) and bake it up to a distance of about 1km from the contact. A granite body like this is called a **pluton** (pronounced **ploo**-ton) from Pluto, Greek god of the underworld. New metamorphic minerals and textures appear in the surrounding baked zone, which is known as a **metamorphic aureole**, a **thermal aureole** or a **contact aureole** (Fig. 1.8). Temperatures are obviously highest right next to the granite and decrease outwards into the cool country rock. A common product of contact metamorphism is a fine-grained, tough and somewhat flinty rock without foliation called **hornfels**. Examples of hornfels derived from shale and basalt are among the rocks described in chapter 2, and

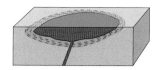

Figure 1.9 Simplified block diagrams of a granite pluton (red) and its metamorphic aureole (orange) shortly after it was intruded (left) and today (right), after the overlying rock and the top of the pluton have been stripped off by erosion.

7

further examples of contact metamorphic rock are discussed in the first section of chapter 4. Contact metamorphism generally happens at depths of a few kilometres so, as with orogenic belts, the land surface must rise and the overlying rocks must become stripped off by erosion for the pluton and its metamorphic aureole to be exposed today (Figs 1.8, 1.9).

1.3.2 Hydrothermal metamorphism

Hydrothermal metamorphism is caused by hot water. It happens in volcanic regions where shallow bodies of underground magma heat the nearby cold **groundwater**. Groundwater is rainwater or seawater that has drained down and filled any voids and crevices beneath the surface. The heated water is less dense than nearby cold water, so it rises (red arrows in Fig. 1.10), flowing through interconnected pores and cracks.

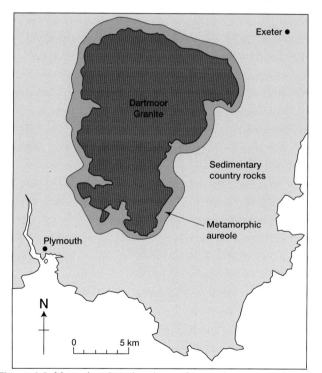

Figure 1.8 Map showing the thermal aureole surrounding the 280 million-year-old pluton of granite exposed at Dartmoor in Devonshire, England. The country rock is composed of sedimentary rocks that were deposited about 400 million years ago. *After R. Mason.*

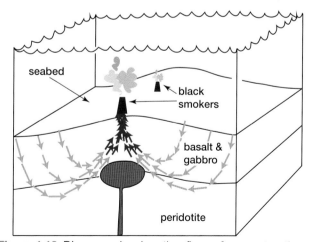

Figure 1.10 Diagram showing the flow of seawater through interconnected fractures in rock beside a sub-oceanic magma chamber at a spreading ridge. For an explanation see the text.

As it moves upwards it is replaced by colder groundwater (blue arrows in Fig. 1.10) in a process known as thermal **convection**. The flowing hot water reacts with the minerals in the rocks through which it passes, creating new hydrous minerals. Hydrothermal metamorphism is particularly important in the vicinity of newly formed basaltic rocks at so-called **spreading ridges** where tectonic plates are moving apart (see Appendix 1, section A1.2). Cameras in submersibles at such places have filmed rising plumes of superheated dark, cloudy water called **black smokers** streaming from vents on the seabed (Fig. 1.11). The uptake of water by basaltic rocks during hydrothermal metamorphism is discussed in the first part of chapter 3, and its implications for volcanism above subduction zones are mentioned in chapter 4.

Figure 1.11 One of the Earth's hottest and deepest black smoker hydrothermal vents which is located on the Mid-Cayman Rise in the Caribbean. The temperature of the expelled H_2O-rich fluid is about 400°C. The 'smoke' (centre of field against black background) consists of dark, fine-grained particles suspended in the fluid. The particles come out of solution as the fluid cools and mixes with cold seawater. Deep-sea shrimps and other life-forms appear to thrive in this bizarre pitch-black environment. *Photo courtesy of Chris German, WHOI/NSF, NASA/ROV Jason 2012©Woods Hole Oceanographic Institution.*

1.3.3 Dynamic metamorphism

Dynamic metamorphism is a localized variety of metamorphism that occurs within major fault zones, and is caused by fault movement. In the top few kilometres of the continental crust it involves the fracturing and comminution of pre-existing rocks, in a process called **cataclasis** (pronounced cata-**clay**-sis). The products are described as cataclastic rocks, or **cataclasites**. Where the faulted rocks are quite cool and shallow they usually end up as a chaotic, loosely-bound mixture of angular broken rock fragments and rock dust called **fault breccia** (pronounced **bretch**-ya). If the product is nearly all pulverized, with few fragments, it is called **fault gouge**. Many geologists do not consider fault breccia and fault gouge to be metamorphic rocks because they are merely loose, unconsolidated mixtures of crushed rock fragments. At greater depth, however, where it is warmer and the grains are pressed more tightly together by the weight of overlying rocks, the faulted rocks often end up as a strong, extremely fine-grained, streaky-looking rock called **mylonite** (pronounced **my**-la-nite; Fig. 1.12). A fault zone at such depths is called a **shear zone**. Shear zones are evolving at depth in the continental crust today wherever major faults are active, such as at plate boundaries (see Appendix 1) and beneath mountain ranges. Various products of dynamic metamorphism, particularly as it affects granite, are described in chapter 2, and the mechanism is discussed in the second part of chapter 3.

1.3.4 Shock metamorphism

Shock metamorphism is an extremely violent process that affects the rocks below the impact site of a giant meteorite. It is a localized and, thankfully, rare phenomenon;

Figure 1.12 Banded and slightly crumpled mylonite from a kilometre-wide, inclined shear zone that can be traced for a great distance immediately to the east of a fault line in NW Scotland known as the Moine Thrust. The Moine Thrust marks the western margin of the Grampian orogenic belt, shown mainly in yellow on the map in the photo.

Figure 1.13 Barringer Meteorite Crater in Northern Arizona. The crater is 1.2km in diameter and was excavated some 50,000 years ago by a huge explosion following the impact of a chunk of iron from space thought to have been travelling at about 20km per second and measuring 50 metres across. This is the best-preserved crater of almost 200 ancient impact craters that have so far been recognized on the Earth's surface. The severe damage caused to the rocks beneath the crater by shock waves from the explosion is known as shock metamorphism. *Shutterstock/ Action Sports Photography.*

such impacts have not been observed in recorded history. The meteorite explodes on striking the ground because its kinetic energy (the energy due to its speed) is instantly turned into intense heat that vaporizes the meteorite and the adjacent rock below the point of impact. The explosion excavates a crater (Fig. 1.13). The rocks below the crater become crushed, shaken, and melted as intense shock waves pass through them. Shock metamorphism is like an extreme kind of dynamic metamorphism. Evidence for it is preserved only rarely on Earth because the surface is continually being refreshed by geological processes, but evidence is widespread on the surfaces of the Moon, Mars and Mercury, and of small bodies in the Solar System. More about shock metamorphism is presented in the last section of chapter 4.

2 The petrography of metamorphic rocks

Petrography is the description of rocks. Describing things – making observations – is the first, and arguably the most important, step in any scientific investigation, so it is a good place to start. This chapter describes a selection of metamorphic rocks, some common, some less so. Most of them are regional metamorphic rocks. A few are products of dynamic and contact metamorphism.

The descriptions are organized by protolith into six sections, starting with metamorphosed sandstone, followed by metamorphosed shale, then limestone, granite, basalt and peridotite. Though not strictly part of petrography, a brief note on the protolith precedes each section, and examples are given, where relevant, of where the rocks might be met in everyday life. The descriptions cover the minerals and texture of each rock. A strong emphasis is placed on the minerals, showing what they look like and stating what they are made of. All two-dozen important minerals listed in Appendix 2 are included.

The descriptions extend to the appearance of rocks in **thin sections**. A thin section is a very thin slice of rock that can be examined using a polarizing microscope. Readers who are not familiar with the polarizing microscope should first read Appendix 3, which explains technical terms used in the descriptions, such as the abbreviations '**XP**' and '**PPL**'. As Appendix 3 points out, anyone can quickly gain experience of rocks in thin sections by clicking on 'collections' at the web-based Virtual Microscope (virtualmicroscope.org) and taking a look for themselves.

Finally, while this chapter is about descriptions of rocks, it does digress a little by commenting on the more obvious aspects of how the rocks were made. However, more theoretical aspects of how they were made are saved for chapter 3 and later.

2.1 Quartzite and metapsammite

2.1.1 Quartzite

Quartzite is a metamorphic rock derived from sandstone that is made entirely, or almost entirely, from grains of quartz.

The name also applies, incidentally, to *unmetamorphosed* quartz-rich sandstone, which closely resembles metamorphic quartzite in appearance.

Quartzite is typically a tough, white or cream-coloured rock with no foliation (Fig. 2.1). The individual quartz grains are usually visible. In the landscape, it stands proud as barren mountain peaks where neighbouring less-resistant rocks have been preferentially stripped away over time by erosion (Fig. 2.2).

The mineral quartz has the formula SiO_2 (silicon dioxide) which is called silica. (All mineral formulae are explained in Appendix 2.) It is mechanically hard and chemically inert, so grains of quartz are great survivors in the successive processes of weathering, erosion, transport and deposition at the Earth's surface. As a result, quartz grains become concentrated in sand, while grains of other minerals are mostly destroyed by mechanical disintegration or by chemical attack.

Figure 2.1 A piece of quartzite similar to the rock from near the summit of Errigal Mountain (Figure 2.2). It consists entirely of quartz. *Shutterstock/sandatlas.org.*

Figure 2.2 Errigal Mountain in County Donegal, Ireland. Its white summit is made of quartzite. *Shutterstock/Patrick Mangan.*

Figure 2.3 A clear, glassy-looking, prismatic crystal of quartz extending from a cluster of interlocking quartz crystals. *Shutterstock/Peter Hermes Furian.*

Quartz is known to mineral collectors for its beautiful clear glassy-looking **crystals** with a prismatic (elongated) shape, hexagonal cross-sections and pointed ends (Fig. 2.3).

The term *crystal* may need clarification. All mineral grains are **crystalline** – that is, they have a regularly repeating internal pattern of atoms called the crystal structure – but few of them are crystals. The word *crystal* is reserved for specimens, like the quartz in Figure 2.3, with flat surfaces (called crystal faces) that have developed naturally during growth, usually in a fluid-filled cavity beneath the ground.

Quartz in quartzite does not exist as crystals but as unevenly shaped grains, because the grains remained packed tightly together as many small ones coalesced to make fewer larger grains during metamorphism. However, geologists often use the word *crystal* in a lax way for any mineral grain, whether it has crystal faces or not.

Quartzite sometimes preserves an original sedimentary feature known as cross-bedding. This provides so-called 'way-up' evidence; it shows the order in which the beds (layers) of sand were deposited (Fig. 2.4), and can help when one is trying to sort out the often complex way in which beds have been folded, and even overturned, during orogenesis and metamorphism.

Figure 2.4 Cross-bedding in quartzite from Ontario, Canada. The layers sloping down to the left are truncated (cut across) by younger layers sloping down to the right from the hammer head. This indicates that the layers of rock are the same way up now as they were when they were deposited.

Figure 2.5 Quartzite is a strong, hard rock, ideal for railway ballast. *Shutterstock/Sigur.*

Since it is such a tough, hard-wearing rock, crushed lumps of quartzite are ideally suited for railway ballast, which forms the bed of stone onto which the tracks are laid, and which must withstand the pounding of heavy locomotives (Fig. 2.5).

2.1.2 Metapsammite

Psammite (from the Greek word *psammos* = sand) is an obsolete word for sandstone, but it is still used when referring to a sandstone protolith, and the name *metapsammite* prevails as the most popular term for metamorphosed sandstone that is not pure quartzite. Some geologists leave out the prefix *meta*, and use *psammite* alone for metamorphosed sandstone.

Psammitic protoliths are dominated by grains of quartz, and usually also contain other ingredients such as mud (in greywacke sandstone) or feldspar (in feldspathic sandstone). Metapsammites are therefore easily recognized because they are made largely of quartz.

If metapsammite is foliated (banded), it can be called psammitic gneiss. An example of high-grade psammitic gneiss in thin section can be seen using the Virtual Microscope (VM) at virtualmicroscope.org. It is sample M06, garnetiferous quartzite, in the VM GeoLab collection. Images of it are shown in Figure 2.6. It consists largely of quartz with a few scattered grains of garnet, and a little feldspar. The quartz is colourless with low relief in PPL, making it quite featureless, but the grains can be seen clearly in XP by their grey and white interference colours. (The reader is reminded that terms, like XP, for minerals in thin sections are fully explained in Appendix 3.) The quartz grains are up to 2cm across (it is instructive

Figure 2.6 Screen shots of GeoLab garnetiferous quartzite M06 in PPL (top) and XP (below), 5mm wide. Quartz dominates; it is colourless in PPL and grey or white in XP. A little garnet (grey blob in PPL; black in XP) and feldspar (thin parallel streaks) are also present.

to look at the entire GeoLab thin section in XP at the lowest magnification). The garnet forms high-relief (greyish in PPL) rounded grains that are isotropic (black in XP). The feldspar forms grains with long, thin parallel outlines. Most of it is potassium feldspar (K-feldspar, $KAlSi_3O_8$; see Appendix 2). K-feldspar is colourless like quartz in PPL. It can be distinguished in XP because it commonly displays criss-crossing grey bands known as **tartan twinning** (Fig. 2.7).

2.2 Metapelite

Metapelite is the preferred name for metamorphosed **shale** and mudstone. These related protoliths are both examples of lithified mud or clay, and are together classified as mudrock.

14

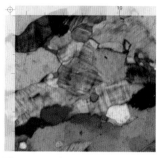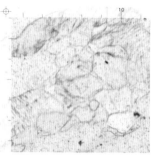

Figure 2.7 A cluster of K-feldspar grains in GeoLab garnetiferous quartzite M06 in XP (left) and PPL, showing tartan twinning. The field is 0.7mm wide. The image can be located, if wished, near the top edge of the GeoLab thin section, 10mm from the left side.

They differ in that shale is fissile, i.e. it splits into thin sheets, called laminations, parallel to the original sedimentary layering, whereas mudstone is not fissile. Just as psammite is an obsolete word for sandstone, so **pelite** (from the Greek word *pelos* = clay) is an obsolete word for mudrock, but it remains widely used, in the context of metamorphism, to mean a mudrock protolith. Indeed, many geologists use the term *pelite* not just for the protolith, but also for metamorphic rocks of pelitic composition. Strictly, however, the latter should not be called pelites but metapelites, and this approach is adopted here.

Mudrock is the most abundant kind of sedimentary rock in the continental crust, and shale is the most abundant kind of mudrock. In the account that follows the name *shale* is widely used, but in all cases it can be switched for the more general name mudrock.

The particles of mud from which shale is made are mostly flakes of **clay minerals** and tiny chips of quartz less than about 0.05mm across. Clay minerals comprise much of the insoluble material that is produced by chemical weathering at the Earth's surface. Several distinct kinds of clay mineral exist, with names such as kaolinite, smectite and illite. They are sheet silicates (see Appendix 2), so they are hydrous, and steam (H_2O) will be released from them if they are baked in a kiln – as happens when clay is fired to make pottery. All clay minerals contain a good deal of the element aluminium (Al) and many also contain some iron (Fe) and magnesium (Mg). The element calcium (Ca) is usually rare. These compositional traits dictate the metamorphic minerals in metapelites.

As noted in chapter 1, metapelites in orogenic belts are commonly foliated, occurring as slate, phyllite, schist and gneiss, with increasing grade. The descriptions of metapelites in the next four subsections are all of foliated types.

2.2.1 Slate

Slate is made from shale by low-grade metamorphism in orogenic belts between about 300°C and 400°C. Slate needs little introduction; it is familiar to everyone as a roofing material (Figs 2.8 and 2.9). It can be split easily into thin, flat sheets with smooth surfaces. The sheets are strong, impermeable to water, and resistant to weathering.

Figure 2.8 Part of a slate roof on the author's house.

Figure 2.9 A slate quarry at Llanberis in North Wales. Large blocks of rock are sawn into small blocks that are split into thin sheets and trimmed to make roofing slates. Quarries in North Wales produced most of the roofing slate used in Britain as its cities expanded in the nineteenth and early twentieth centuries. *Shutterstock/Gail Johnson.*

The tendency for slate to split into sheets is called **slaty cleavage**. It occurs because abundant tiny plate-like flakes of the mineral **muscovite** in the rock lie perfectly parallel to each other. When the rock is split the fracture propagates in the direction of the aligned muscovite plates. Moreover, each individual plate is prone to being split internally into even thinner sheets. This feature of muscovite is also known as **cleavage**; it is called *mineral cleavage* to distinguish it from the rock's slaty cleavage. Cleavage in muscovite is illustrated in Figure 2.10 where a large sheet of muscovite is being prised open along its cleavage planes, using the tip of a penknife blade, to give transparent, flexible sheets that look like cellophane. Muscovite, like the clay minerals in the original shale, is an aluminium-rich sheet silicate. Its chemical formula, $KAl_3Si_3O_{10}(OH)_2$, looks complicated but is simple enough when understood in terms of its crystal structure, as is explained in Appendix 2.

The muscovite plates in slate cannot be seen with the naked eye because they are far too small, but they can be seen using a **scanning electron microscope**, abbreviated SEM (Fig. 2.11). An SEM is a powerful imaging tool, used in many branches of science. The way it works, and the different kinds of pictures it can produce, are described in Appendix 4.

Slate occurs in low-grade metamorphic belts called **slate belts**. An example is the Carolina slate belt in the eastern USA, and another is the Cambrian slate belt of North Wales, where the Llanberis quarry shown in Figure 2.9 is located. The orientation of slaty cleavage within these belts is usually steeply inclined, at right angles to the directed stress that squeezed the rocks during orogenesis.

In some outcrops of slate, the original sedimentary **beds** (layers) are visible. Where the beds have been folded, the slaty cleavage cuts them at different angles in different parts of a fold, as shown in Figure 2.12. Here, 'W-shaped' folds and slaty cleavage developed together in response to horizontal directed stress, indicated by the opposing arrows. The slate sample shown in Figure 2.13 comes from a place where the cleavage and bedding are nearly at right angles to each other, like the one marked 'A' in Figure 2.12.

Slate can be of different colours. Green slate (as in Fig. 2.13) contains the green sheet silicate mineral, **chlorite**, which commonly accompanies muscovite. Chlorite also has mineral cleavage (see Fig. 2.16). Purple slate from North Wales takes its colour from traces of the accessory mineral, **hematite** (red iron oxide), with the chlorite. Some purple slates have

Figure 2.10 A large sheet of muscovite being prised open with a penknife blade along its cleavage to produce thin, flexible transparent sheets.

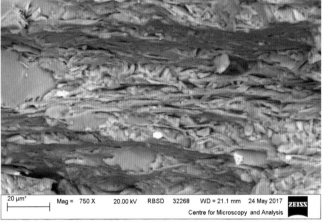

20 µm   Mag = 750 X   20.00 kV   RBSD   32268   WD = 21.1 mm   24 May 2017   ZEISS
Centre for Microscopy and Analysis

Figure 2.11 The broken edge of a piece of slate magnified using a scanning electron microscope. The alignment of tiny plates of muscovite is clearly visible. The field of view is 0.12mm wide.

sporadic oval green coloured patches on their surfaces (Fig. 2.14). These are called **reduction spots** and occur in places where the hematite has been removed by natural chemical reduction, leaving just the green chlorite to provide the colour.

Although the flakes of muscovite and chlorite in slate are too small to see, these minerals can be identified reliably using an analytical technique known as **X-ray powder diffraction (XRD)**. XRD can be used for identifying minerals in any rock, fine-grained or not. It is described in Appendix 4, where an example of an XRD 'spectrum' from a sample of slate is illustrated (Fig. A4.6).

XRD can also be used to infer the extent to which shale has been turned into slate. One of the clay minerals in shale is called **illite**. Illite is a poorly formed kind of muscovite. On an XRD spectrum, illite has rather broad peaks, in contrast to the

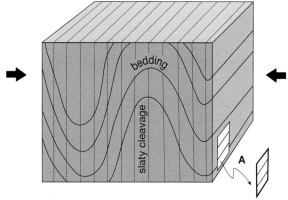

Figure 2.12 Diagram of a metre-scale block of slate showing the relationship between slaty cleavage and folded bedding (sedimentary layering). For an explanation see the text.

Figure 2.13 A close-up view of the edge and surface of a roofing slate, about 5mm thick, from Killaloe in Ireland. The original sedimentary beds are thin green-grey and pale creamy green laminations. On the edge the beds, which are nearly horizontal, can be seen to be sliced across by the slaty cleavage, which is vertical. It corresponds to the position labelled 'A' in Figure 2.12.

sharp ones seen in muscovite. As metamorphism progresses, illite gradually changes to proper muscovite and its peaks become steadily sharper. The sharpness increases with metamorphic grade, and can be expressed quantitatively as the

Figure 2.14 A green reduction spot in a purple-grey slate. *Photo courtesy of Steven Dutch.*

Kübler Illite Crystallinity Index. This is the width of the tallest peak (the so-called (001) peak of illite) measured in degrees at half the full peak height. The (001) peak is at 9° in Fig. A4.7. Starting from shale, the peak's width shrinks from over 1.4° down to 0.25° with increasing temperature of diagenesis. When the Kübler index is between 0.42° and 0.25° conditions are sometimes referred to as 'very low-grade metamorphism', or anchizone metamorphism. The anchizone is thought to relate to temperatures between about 200°C and 300°C, as is shown in Figure 1.6. Slate can form in the anchizone as well as under low-grade metamorphic conditions.

As a final comment, not all slate is produced from shale. It can also be made from other protoliths, for example, from volcanic ash.

2.2.2 Phyllite and low-grade schist

Pelitic rocks metamorphosed at low grade, but above the temperature where slate is formed, tend to split into slightly wavy sheets (Fig. 2.15) with a silky smooth surface and a silvery or golden sheen. Such rocks are called **phyllite** (from the Greek word *phyllon* = leaf). The minerals are typically the same as in slate – quartz, muscovite and chlorite – but the grains are larger, and just about visible with the aid of a hand lens.

With increasing grade, the grains get large enough to be seen with the unaided eye, and the rock is then called schist instead of phyllite. Pelitic schist always contains one kind, and often both kinds, of **mica** – muscovite and **biotite**. Biotite has cleavage like muscovite, but is black (Fig. 2.16B). It differs in composition from muscovite by containing iron (Fe) and magnesium (Mg) instead of aluminium (Al). Its formula is

Figure 2.15 A sample of phyllite about 10cm wide from Ashleam Bay, Achill Island, western Ireland.

Figure 2.16 (**A**) cleavage flakes of the green sheet silicate chlorite; (**B**) biotite (black mica) showing how it can be split along cleavage planes in the same manner as muscovite.

$K(Mg,Fe)_3Si_3AlO_{10}(OH)_2$, and, like that of muscovite, is linked to the crystal structure (see Appendix 2).

The potential for schist and phyllite to split along the direction of the aligned mica plates is called **schistosity**. Schistosity is really the same thing as slaty cleavage, but the grains are larger, and the split surface of the rock is not flat but undulating.

The undulating surface of schist and phyllite is sometimes wrinkled, rather than smooth, because the schistosity has been crumpled into tiny, evenly spaced folds like a miniature corrugated roof (Fig. 2.17). The wrinkles are called **crenulations**, and they form when the local stress in the orogenic root changes to a new direction (see Fig. 2.18). If the squeezing is intense, the limbs of the crenulations may become aligned, giving rise to a new direction of schistosity parallel to those limbs.

2.2.3 Minerals and textures of medium-grade schist

Four minerals that can appear with mica in medium-grade pelitic schist are garnet, staurolite (pronounced **stor**-a-lite), kyanite (pronounced **kye**-a-nite) and sillimanite. They have distinctive appearances as crystals, which are shown in Figure 2.19.

Garnet in metapelites is an iron-aluminium silicate with limited **atomic substitution** of iron by the elements calcium (Ca), magnesium (Mg) and manganese (Mn). Atomic

Figure 2.17 Specimen of mica schist with crenulated schistosity.

17

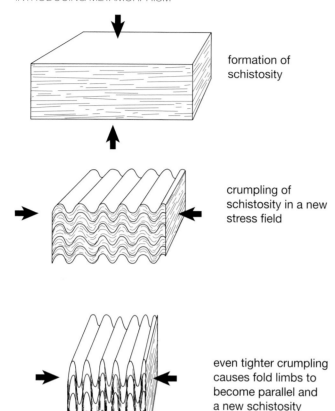

Figure 2.18 Diagrams showing an original schistosity in phyllite or schist (top), and how crenulations result from crumpling of the schistosity (middle) by a new directed stress (black arrows). With intense compression, a new schistosity will develop (bottom).

formation of schistosity

crumpling of schistosity in a new stress field

even tighter crumpling causes fold limbs to become parallel and a new schistosity is created

substitution is important in many metamorphic minerals and is explained in Appendix 2. The iron **end-member** is called **almandine** (pronounced **ah**-mun-deen) and its formula is $Fe_3Al_2(SiO_4)_3$ (see Appendix 2). It is hard and dense, and it often occurs as beautiful, red, glassy-looking crystals with twelve diamond-shaped faces. The shape is known as a **rhombic dodecahedron** (Fig. 2.19A). Crystals, like this, with good crystal faces are described as **euhedral** (pronounced you-**heed**-ral) from the Greek meaning 'good faces'.

Figure 2.19 Four minerals from medium-grade metapelites. (**A**) A garnet crystal with the shape known as a rhombic dodecahedron which, in Greek, means twelve faces each with the shape of a rhombus (diamond). (**B**) A cruciform (cross-shaped) twin crystal of staurolite. (**C**) An elongate blue, blade-shaped prism of kyanite. (**D**) A cluster of grey-white prisms of sillimanite in a pelitic hornfels.

Grains of garnet in schist, whether euhedral or not, are often much larger than the grains around them. Such grains are called **porphyroblasts** (pronounced: **poor**-fi-roh-blast). Porphyroblasts of garnet are of interest because their relationship to the schistosity shows whether they grew at an early stage in the rock's history, or at a late stage. If the garnet grew early, the mica plates are bent where the directed stress has pushed them against the porphyroblasts, which act like rigid pellets. The pattern of the mica schistosity, deflected around the garnets, resembles eyes peeking from the rock (Fig. 2.20). The garnet 'eyes' are called **augen** (pronounced **ow**-gen; ow as in cow, and a hard 'g') which is the German word for eyes.

An example of garnet augen seen through a microscope can be examined in the Virtual Microscope sample GeoLab, M09, which is a garnet-mica schist. This rock is illustrated and

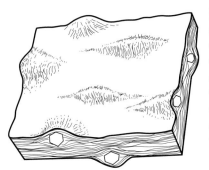

Figure 2.20 Sketch of a piece of schist with augen of garnet. Garnet porphyroblasts (large crystals) grew first, and the schistosity was later moulded around them by flattening.

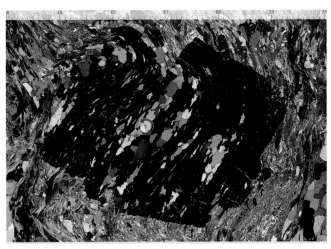

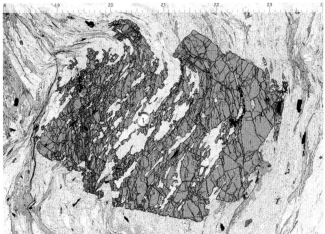

Figure 2.21 GeoLab garnet-mica schist M08 in XP (top) and PPL. Garnet forms the single large porphyroblast that is isotropic in XP. It grew *late*, after the schistosity had been crumpled, and trapped some of the folded layers of quartz. The field of view is 6mm wide.

described fully in Appendix 3 (Fig. A3.9) where the rock has been selected to introduce the appearance in a thin section of five minerals common in metapelites. These are garnet, muscovite, biotite, quartz, and a mineral called plagioclase (pronounced **plaj**-ee-oh-claze) which is a variety of feldspar (see Appendix 2). The garnet grains in M09 (Appendix 3) occur as augen, and they should now be examined online at virtualmicroscope.org.

A very different texture is seen where garnet porphyroblasts grew late, as they did in VM GeoLab garnet-mica schist M08. In thin section (Fig. 2.21) a single garnet porphyroblast (high relief in PPL and black (isotropic) in XP) at rotation 1 clearly grew after an initial schistosity had already been crumpled into folds. The crumpled schistosity comprises thin layers of quartz (clear in PPL, variably grey in XP) alternating with layers of muscovite (bright colours in XP). The growing garnet replaced the existing muscovite in its path, but did not replace all the quartz, some of which became engulfed as so-called **inclusions**. The curved trails made by the inclusions are part of the folded schistosity outside the garnet. A porphyroblast full of inclusions like this is known as a **poikiloblast** (pronounced **poy**-kill-oh-blast). The rock also contains chlorite. A clump of chlorite at rotation 3 is seen online to be **pleochroic** from green to almost colourless.

This same rock, M08, was selected to demonstrate the powerful technique of scanning electron microscopy (SEM) which is outlined in Appendix 4. Figure A4.1 shows various images of the same area of M08, including a PPL image of a thin section, a back-scattered electron image, and various X-ray element maps.

Leaving garnet, the other three minerals named above, staurolite, kyanite and sillimanite, also commonly occur as euhedral crystals (see Fig. 2.19 B to D). Staurolite normally forms brown prisms (as in the schist on the front cover of the book), but in Fig. 2.19B it forms a so-called cruciform (cross-shaped) **twinned crystal**. **Kyanite** is typically blade-shaped and blue with cleavage (Fig. 2.19C). **Sillimanite** forms spiky white prisms (Fig. 2.19D). All three minerals are aluminium-rich, consistent with a shale protolith. Kyanite and sillimanite both have the formula Al_2SiO_5. **Staurolite**'s formula is complex, approximating two lots of Al_2SiO_5 and one of $FeO(OH)$.

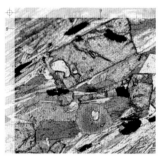
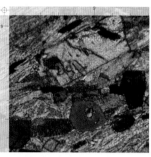

Figure 2.22 Enlargement of part of the left side of Figure 2.25 in PPL (left) and XP showing a rectangular crystal of staurolite (pale yellow in PPL). The green mineral below it, incidentally, is an end-on view of a prism of the accessory mineral, **tourmaline** (see appendices 2 and 3).

Figure 2.23 Screen shot of kyanite from kyanite schist (VM OU collection number S339 24) in PPL (left) and XP. Other minerals present are garnet (lower left), muscovite (blue in XP), biotite (brown in PPL), and quartz. The field is 1.5mm wide.

How do these last three minerals appear in thin section? Staurolite can be seen in VM GeoLab, M10, a garnet-mica schist from Connemara in western Ireland. This beautiful rock contains staurolite, garnet, muscovite, biotite, quartz, and plagioclase. Staurolite in PPL is pale yellow (Fig. 2.22) and pleochroic (check rotation 1). It has high relief, it lacks cleavage, and it tends to have rectangular outlines. In XP it has grey to cream interference colours and parallel **extinction**.

Examples of kyanite and sillimanite in thin section are shown, respectively, in Figures 2.23 and 2.24. Both minerals in PPL are colourless with high relief and elongate outlines. In XP they are both grey, white or creamy yellow (first order interference colours, though sillimanite can be purple or blue) with parallel extinction. Their cleavage distinguishes them when it is seen end-on, as in the images here. Kyanite blades have two cleavages, one more prominent than the other, at an angle of about 100° (Fig. 2.23), while sillimanite has square outlines with a single diagonal cleavage (Fig. 2.24). In addition, sillimanite commonly develops as fibres, looking like bundles of hair (Fig. 2.24). Fibrous sillimanite and prismatic sillimanite can occur together.

Before leaving this account of minerals in metapelites, the staurolite-bearing schist, M10, has garnet porphyroblasts that are worth taking a closer look at. They have inclusion trails and are also augened by the schistosity, as is evident in Figure 2.25 and is best seen online. Moreover, a prism of staurolite, midway between the two garnet augen in Figure 2.25, cuts across the schistosity, showing that it grew last of all.

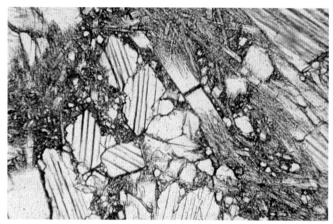

Figure 2.24 A bundle of sillimanite prisms in PPL in a sillimanite rock. They are seen 'end-on' to the left of the centre, where they show characteristic square cross-sections with a single diagonal cleavage. Elsewhere the sillimanite is seen as prisms on their sides and as fibres. Width of field is 1.5mm.

Now some garnet porphyroblasts in this schist (but unfortunately not in the VM M10 thin section) contain *folded* inclusion trails. Adding this detail, the full history of grain growth and deformation is summarized in the sketch and caption in Figure 2.26.

2.2.4 The regional distribution of minerals in low- and medium-grade schist

Leaving thin sections aside, and returning to field outcrops, the geographical distribution of the minerals in pelitic schist

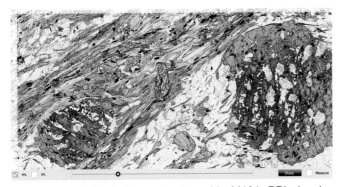

Figure 2.25 VM GeoLab garnet-mica schist M10 in PPL showing two garnet porphyroblasts augened by the mica's schistosity. For a description see the text. Width of field is 8mm.

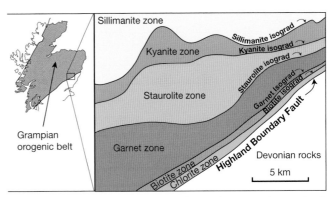

Figure 2.27 Map showing isograds and zones in part of the Grampian orogenic belt in north-east Scotland. For an explanation see the text.

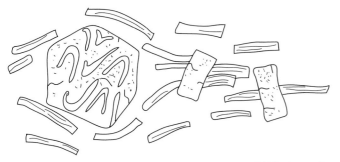

Figure 2.26 Sketch summarizing the texture in M10. An early schistosity was crenulated and the small folds became trapped inside a garnet porphyroblast that grew later. Later still, under a different direction of stress, a new mica schistosity developed and the garnet porphyroblast became augened. Finally, rectangular grains of staurolite grew randomly across the schistosity, in the absence of directed stress.

shows a pattern in many metamorphic belts that is thought to reflect a regional variation in the grade (temperature) of metamorphism. The pattern is seen, for example, in the part of the Grampian orogenic belt in NE Scotland shown in Figure 2.27. Here, low-grade metapelites (slate and phyllite) near the margin of the belt (the Highland Boundary Fault) are greenish in colour and contain chlorite and muscovite. Travelling northwards, phyllite grades into mica schist and new minerals join the muscovite and chlorite in a sequential way. First the schist darkens in colour as biotite joins in, then further on crystals of red garnet are added, and after that staurolite, then kyanite and finally sillimanite. These changes accompany increasing grade and are described as **prograde** metamorphic changes.

The six minerals – chlorite, biotite, garnet, staurolite, kyanite and sillimanite – are known as **index minerals**. Lines have been traced on the ground, and plotted on a map (Fig. 2.27) to delineate the first appearance of the index minerals. The lines are named **isograds** because they are thought to connect rocks of equal grade. Each isograd is named by the index mineral that defines it. The area of ground between one isograd and the next is called a **metamorphic zone**. A zone takes its name from the new index mineral within it. For example, the garnet zone is the tract of land between the garnet isograd and the staurolite isograd, and is shown in red in Figure 2.27.

The metamorphic zones in Figure 2.27 were discovered in the late 1800s by George Barrow, an officer of the British Geological Survey, while preparing a geological map of this part of the Scottish Highlands. Since then, similar zones have been found in many other metamorphic belts throughout the world. They are called **Barrow zones** in honour of George Barrow.

The names of metamorphic zones are used to give a general idea of metamorphic temperature, as was briefly noted in chapter 1. Zones are more refined than *grades*. The chlorite and biotite zones fall within low-grade metamorphism, while the other four zones subdivide medium-grade conditions.

The ranges of the various minerals in metapelites through the six zones, and beyond, are summarized in Figure 2.28. Low-grade metapelites start out with quartz, muscovite,

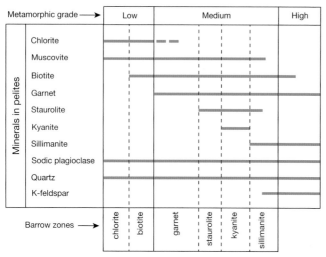

Metamorphic grade ⟶	Low	Medium		High

Minerals in pelites:
- Chlorite
- Muscovite
- Biotite
- Garnet
- Staurolite
- Kyanite
- Sillimanite
- Sodic plagioclase
- Quartz
- K-feldspar

Barrow zones ⟶ chlorite | biotite | garnet | staurolite | kyanite | sillimanite

Figure 2.28 The ranges, with increasing grade through the Barrow zones, of minerals in metapelites.

22

chlorite and sodic (Na) plagioclase, and, as the grade increases, the other five index minerals are added sequentially. Once introduced, the index minerals mostly continue into the next zone, and even beyond it. Chlorite disappears once garnet gets established. Muscovite and staurolite disappear in the middle of the sillimanite zone, where K-feldspar appears as a new mineral. Kyanite, however, only exists in the kyanite zone. It changes to sillimanite once the sillimanite isograd is crossed. The diagram also shows what happens at high grade. Biotite eventually disappears, but quartz, Na-plagioclase, K-feldspar, garnet and sillimanite continue on to the end. An example of a high-grade metapelite is described in chapter 5.

2.2.5 Pelitic gneiss and migmatite

Metapelites formed in the sillimanite zone beyond where muscovite disappears carry a much reduced amount of mica (just biotite), a good deal of quartz and feldspar, and garnet and sillimanite. With reduced mica, they may lose their schistosity and become pelitic **gneiss**. Gneiss is a widely used, but poorly defined, name. It usually refers to rocks with a coarse grain size (e.g. > 1mm) and a banded appearance with some alignment of elongate grains, but with no particular tendency to be split like schist. Gneiss can be derived from protoliths other than shale. While pelitic gneiss is common, psammitic gneisses are also common, as are **semipelitic** gneisses (having a composition transitional between normal

pelitic and psammitic compositions). Also, many kinds of gneiss are derived from igneous protoliths.

Staying with pelitic gneiss, this rock sometimes has a distinctive appearance with dark, biotite-rich layers alternating, or mixed more irregularly, with light-coloured layers or lens-shaped patches composed of quartz and feldspar. Folding of the layers may be complex. Rocks with this appearance are named migmatitic gneiss, or simply **migmatite**. Migmatite means 'mixed rock' (from the Greek word *migma* = mixture). An example of folded migmatite from the island of Naxos in Greece is the background on the cover of this book. A separate example from Norway, with intricate folding, is shown in Figure 2.29.

Migmatites have their own terminology. The light-coloured part is named the **leucosome** (pronounced **loo**-koh-zohm) while the adjacent darker material is called the **melanosome** (pronounced **mell**-an-oh-zohm).

Migmatite is widely thought to be a product of **anatexis** (partial melting) of pelitic rock. The leucosome is thought to be made of locally produced, now frozen, granitic magma, while the melanosome is interpreted as the residual rock, left behind after the granite magma has leaked out of it. Melanosome is thus described as a 'melt residue'; it is also called **restite**.

Some definitions restrict metamorphism to the solid state, so technically they imply that migmatite formation is not a metamorphic process. This view is not adopted here. Many high-grade metamorphic rocks are undoubtedly melt residues.

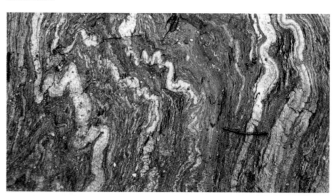

Figure 2.29 Complex folds in migmatite from Geirangerfjord in western Norway. The width of the field is probably about 1 metre. *Wikimedia Commons File:Migma ss 2006.jpg under 'migmatite' – taken by Siim Sepp.*

2.2.6 Metapelite in a contact aureole

The focus turns now from orogenic belts to contact aureoles. Here a common kind of rock is hornfels, a tough, non-foliated, fine-grained rock of any composition (see chapter 1, section 1.4). One particularly distinctive metapelite formed in this setting is called andalusite-cordierite-hornfels, and an example is illustrated in Figs 2.30 and 2.31.

Figure 2.30 A specimen of andalusite-cordierite hornfels that formed from slate in the aureole of a large granite pluton at Skiddaw in the English Lake District. Random white 'matchsticks' are andalusite prisms. Cordierite forms barely perceptible greyish spots. *Photo kindly supplied by John Waring http:// www.negs.org.uk.*

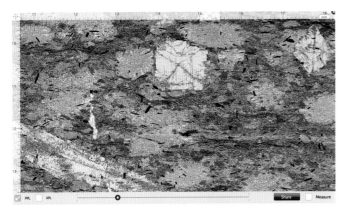

Figure 2.31 Andalusite-cordierite hornfels in PPL. It is derived from slate (Virtual Microscope Leeds collection, named 'slate'). Width of field is 7.5mm. It is described in the text.

Andalusite (pronounced and-a-**loos**-ite) and cordierite (pronounced **cord**-ear-ite) are two aluminium-rich minerals not met so far. The andalusite in Figure 2.30 forms prismatic crystals resembling randomly orientated matchsticks. It has exactly the same chemical formula as kyanite and sillimanite, i.e. Al_2SiO_5. Different minerals with the same chemical composition are said to be **polymorphs** of that composition. Thus kyanite, sillimanite and andalusite are polymorphs of Al_2SiO_5.

Cordierite occurs as porphyroblasts which, in Figure 2.30, are rounded dark 'splodges' that blend into the purplish fine-grained background. Cordierite's formula is $(Mg,Fe)_2Al_4Si_5O_{18}$ (see Appendix 2).

A pelitic hornfels from the Lake District similar to the one in Figure 2.30 can be examined in the VM Leeds collection (it is called 'slate' which is the name of its protolith). It is shown in PPL in Figure 2.31. Here, two large andalusite prisms are seen 'end-on' with square cross-sections, and one is side-on with an elongate outline (lower left). End-on sections have a distinctive diagonal cross-shaped pattern of dark inclusions, which distinguishes andalusite from kyanite and sillimanite. In XP (seen online) it has grey colours and parallel extinction. Cordierite in PPL has low relief and, in XP, it has grey or white interference colours. It might perhaps be mistaken for quartz or feldspar, but it is distinctive enough in this rock because it forms large oval poikiloblasts (inclusion-rich porphyroblasts). In XP (seen online) some of these show sector twinning, looking a bit like a round cake cut into wedges, in different shades of grey. All other grains are tiny. The brown, randomly orientated grains in PPL are of biotite. The small, scattered opaque grains are probably graphite, which is a form of carbon.

The random orientation of andalusite and biotite means hese two minerals interlock to make the hornfels tough and hard to break. The texture suggests that there was no directed stress when the minerals were growing. A stress-free environment is common in aureoles because directed stress cannot be sustained by a body of magma; magma flows, instantly dissipating any directed stress.

2.2.7 The significance of Al_2SiO_5 for inferring metamorphic conditions

Kyanite, andalusite and sillimanite, the three polymorphs of Al_2SiO_5, are each stable within their own specific range of pressure (P) and temperature (T). If a rock contains kyanite,

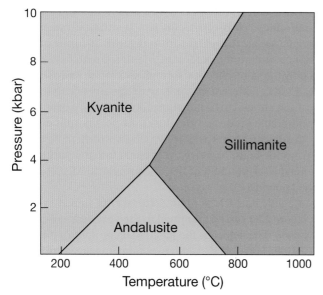

Figure 2.32 Pressure–temperature diagram for Al_2SiO_5 showing the stability fields of kyanite, andalusite and sillimanite.

for example, then it is presumed to have been metamorphosed under P–T conditions within the blue area in Figure 2.32, called the kyanite stability field. The meaning of *stability*, and the way in which the three stability fields were determined, will be explained in chapter 3. For the present, it can be noted that the Al_2SiO_5 diagram has an obvious bearing on interpreting metapelites. For example, it shows that the pressure during *regional* metamorphism (no andalusite) must have been greater than about 4kbar because the kyanite zone is next to the sillimanite zone. It also shows that andalusite hornfels must have been formed at pressures less than about 4kbar.

2.3 Marble

Metamorphosed limestone is known as **marble**. The limestone protolith is almost always made from the calcium carbonate ($CaCO_3$) skeletal remains of planktonic marine organisms, together with shells from larger creatures, which accumulate on the seabed before becoming buried and lithified to make hard limestone. The organisms extract calcium and CO_2 from seawater where they are in solution (calcium is a soluble product of chemical weathering, and is carried by rivers to the sea).

Incidentally, this process, operating over geological time, has continuously removed CO_2 from seawater, increased the amount of limestone in the continental crust, and kept CO_2 in the atmosphere at a steady low level. Today the rate of removal of CO_2 seems unable to keep pace with the rapid anthropogenic production of CO_2, and the level of atmospheric CO_2 is rising alarmingly year on year.

2.3.1 Pure calcite marble

The simplest kind of marble is made entirely of the mineral **calcite** ($CaCO_3$). Pure calcite marble is usually white or pale grey in colour, and when it is crushed it breaks into blocky pieces with no foliation, so at first sight it can resemble quartzite. However, it is much softer and more easily abraded than quartzite; crushed marble would be of little use as railway ballast. Its softness means that marble can be scratched, scraped and chiselled with ease, making it an excellent choice of stone for carving. It has been used throughout history in sculpture and as a decorative stone in buildings. Notable in this regard is the white marble from the quarries at Carrara, perched on the mountain slopes high above the Ligurian Sea in Tuscany, northern Italy (Fig. 2.33). Michelangelo created his masterpiece *David* (Fig. 2.34) from a carefully chosen block of Carrara marble.

Figure 2.33 Marble quarry at Carrara, Tuscany, Italy, looking west. The Ligurian Sea is in the far distance. *Shutterstock/ Alessandro Colle.*

Figure 2.34 Michelangelo's famous masterpiece *David*. It was carved from a block of marble from Carrara. *Shutterstock/ javi_indy.*

Figure 2.35 A block of clear, transparent calcite obtained by splitting a larger crystal of calcite along cleavage planes in three directions. Clear calcite like this is called Iceland spar. *Shutterstock/Eduardo Estellez.*

Marble is amenable to being carved because of the crystalline properties of calcite. Examination of a large crystal of calcite shows that it can be split easily along internal cleavage planes, but instead of having a single direction of cleavage, like muscovite, calcite has three cleavage directions (Fig. 2.35).

Every grain of calcite in marble has these three directions of cleavage within it so, when the rock is being carved, it will fracture easily in any direction chosen by the sculptor. Also, a freshly broken surface of a marble specimen will glitter as it is turned and the light catches the tiny, perfectly flat, mirror-like surfaces where individual grains have parted along a cleavage plane. The grains in marble, and Carrara marble is no exception, tend to be evenly sized. This texture is sometimes described as saccharoidal (sugar-like). Another term used for even-grained rocks like this is **granoblastic**.

Marble is not only easy to carve, but it readily takes a shine when buffed with a polishing paste. Polished marble tiles and worktops have become something of a fashion statement in people's homes, but unwitting house-proud home owners may find themselves dismayed when they discover, too late, that polished marble, being so soft, is very easily scratched! Worse still, the polished surface of marble is easily etched and stained by weak acids like vinegar and lemon juice. Such acids react spontaneously with the calcite grains in the marble, dissolving them.

Geologists take advantage of this chemical reaction with acid when trying to distinguish marble (or its limestone protolith) from rocks that contain no calcite. The surface of the rock is wetted with a drop of dilute hydrochloric acid from a dropper bottle. If the rock contains calcite it will effervesce vigorously as CO_2 gas is released (Fig. 2.36).

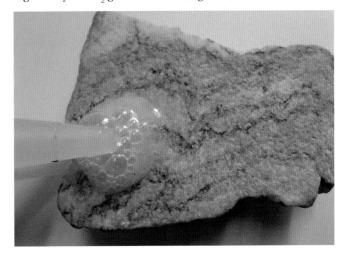

Figure 2.36 Effervescence of CO_2 gas as a drop of dilute hydrochloric acid wets a piece of marble.

The chemical reaction that takes place is written:

$$CaCO_3 + 2HCl = CaCl_2 + H_2O + CO_2$$

calcite　　hydrochloric acid　　calcium chloride　　water　　carbon dioxide

It may come as a surprise to learn that some pure marbles are not made of calcite. Calcium carbonate also exists in nature in another crystalline form or polymorph, named **aragonite**. Aragonite has a higher density than calcite, and in exceptional metamorphic settings where the pressure is abnormally high (as will be reviewed in chapter 4), marble can be made from aragonite. In fact, many shells of marine organisms are made from aragonite rather than calcite. However, aragonite in shells, and in aragonite marble, is not as stable as calcite at the Earth's surface, and it is nearly always in the process of changing slowly into calcite.

2.3.2 Impure marble

Some varieties of limestone contain the calcium–magnesium carbonate mineral named **dolomite**. Such limestone becomes dolomitic marble when it is metamorphosed. Dolomite contains equal numbers of calcium and magnesium atoms; its formula is $CaMg(CO_3)_2$. Compared with calcite, it is equally soft, but is usually cream coloured rather than white, and it does not react vigorously with dilute hydrochloric acid unless it has first been crushed to powder. Dolomite is formed by the natural replacement of exactly half the calcium atoms in calcite by magnesium atoms from seawater during a process called dolomitization. The process takes place on, or just beneath, the seabed in tropical latitudes where calcite is accumulating (e.g. as shells) and seawater is evaporating. It is an example of **diagenesis**, the name for changes that happen to rocks below the temperatures normally associated with metamorphism, as was outlined in chapter 1.

Dolomitic limestone may contain silica (SiO_2), for example as quartz sand grains or as the variety of silica known as opal, which forms the skeletons of certain planktonic organisms like radiolaria and diatoms. After such limestone has been metamorphosed, the resulting dolomitic marble will normally contain magnesium-bearing silicates like diopside and forsterite. Both minerals are introduced in Appendix 2. **Diopside** is a white mineral (see Fig. 2.37) with good cleavage, whose formula is $CaMgSi_2O_6$. It is an example of a kind of silicate known as clinopyroxene. **Forsterite** is green; its formula is Mg_2SiO_4. It is a variety of

olivine, a mineral that is described in Appendix 1 as the main constituent of the Earth's mantle. Forsterite is the so-called magnesian end-member of olivine, as explained in Appendix 2.

Diopside is formed by a simple chemical reaction between dolomite and quartz:

$$CaMg(CO_3)_2 + 2SiO_2 = CaMgSi_2O_6 + 2CO_2$$

dolomite　　silicon dioxide (quartz)　　diopside　　carbon dioxide

Forsterite is formed by a related reaction:

$$2CaMg(CO_3)_2 + SiO_2 = 2CaCO_3 + Mg_2SiO_4 + 2CO_2$$

dolomite　　quartz　　calcite　　forsterite　　carbon dioxide

An example of forsterite-diopside marble in the field is illustrated in Figure 2.37, and a thin section of the same rock is shown in Figure 2.38.

The thin section photographs (Fig. 2.38) are screen shots of VM GeoLab M05, forsterite marble. The rock is well worth exploring online. Calcite and dolomite, the two carbonate minerals, look alike; they are both colourless with high relief. However, they can be distinguished here because the surface of the thin section was treated with a pink dye, called alizarin red S, before the glass cover

Figure 2.37 A rock outcrop of forsterite-diopside-marble at Glenelg in NW Scotland. The forsterite forms small, rusty-looking granules protruding from the surface. The diopside forms white protruding grains, and it also forms the large, rounded white lump on the left. The silicates protrude because they do not dissolve as easily as the calcite and dolomite during weathering.

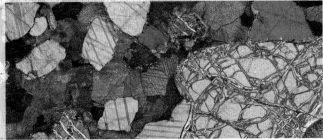

Figure 2.38 Screen shots of VM GeoLab forsterite marble, M05, in PPL (upper image) and XP. The pink colour shows where calcite has been stained. Sets of straight parallel bands are deformation twin lamellae. Forsterite (Mg-olivine) lower right is traversed by narrow veins of serpentine. The width of the image is 5mm.

slip was attached (for an account of how thin sections are made, see Appendix 3). The dye stained the calcite but did not affect the dolomite. Online examination will show how half the thin section was treated in this way. Both minerals commonly display one, two or even three sets of narrow, parallel, coloured bands that show up in XP, and may even be visible in PPL. These are known as **deformation twin lamellae**. They were created when the marble was subjected to stress and became strained. The interference colour of calcite and dolomite in XP is creamy white (often with a hint of twinkly pinks and greens), known as high-order white (see Appendix 3). The colour remains unchanged when a special filter, called a sensitive-tint plate, is inserted into the light path in the microscope. This test allows carbonates to be identified; nearly all other minerals change colour in XP when the sensitive-tint plate is inserted. Unfortunately, this useful test does not currently feature in the Virtual Microscope.

Forsterite occurs as colourless, high-relief grains with rounded outlines. It has bright interference colours in XP and any grains with roughly rectangular shapes tend to have parallel extinction. Each grain is traversed by a network of narrow channels, called veins, filled by a low-relief mineral with grey interference colours. This is the mineral **serpentine**, a water-rich sheet silicate mineral whose formula is $Mg_6Si_4O_{10}(OH)_8$ (see Appendix 2). It formed here at some late stage in the marble's history, when water seeped in and made its way along cracks in the forsterite, converting it to serpentine. The reaction is an example of what is known as **retrograde** alteration. In a few cases the alteration has progressed to the point where no forsterite (or very little) remains (Fig. 2.39, upper pair of images) and the resulting lump of serpentine, where once there had been forsterite, is known as a **pseudomorph** of serpentine after forsterite.

Diopside has the same high relief and second order interference colours as forsterite. It can be distinguished from forsterite because it does not have veins of serpentine, and it may show parallel cleavage cracks (Fig. 2.39, lower pair of images). Grains with cleavage have a very high extinction angle (see Appendix 2), i.e. they go into extinction when the cleavage lines are typically between 30° and 45° from the nearest cross hair.

The above marble was formed under high-grade conditions. At medium grade a different magnesium-silicate mineral, **tremolite**, generally occurs instead of forsterite and diopside, giving tremolite marble. It forms long white prisms (Fig. 2.40). Tremolite has the formula $Ca_2Mg_5Si_8O_{22}(OH)_2$, and is the simplest example of a clinoamphibole. Although its formula looks complicated, it is easy to remember when it is related to the crystal structure of tremolite (see section A2.4.3 of Appendix 2).

Some impure marbles occur as serpentine marble. The serpentine may have formed by the hydration of former olivine (as in Fig. 2.39) but, if water were present during the original metamorphism of the siliceous dolomitic limestone, then the serpentine may have formed directly, as a primary metamorphic mineral. Green serpentine marble has for many years been quarried and carved for use as an ornamental stone (Fig. 2.41).

Limestone may be rendered impure not only by having an admixture of quartz, but also as a result of containing a high proportion of clay minerals (mud), giving it a

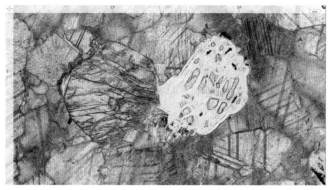

Figure 2.39 Pairs of images in PPL (left side) and XP, each about 1mm wide from forsterite marble, M05. *Top* Forsterite almost entirely replaced by serpentine, making a so-called pseudomorph of serpentine after forsterite (right of centre). *Bottom* Diopside with a hint of cleavage and no alteration to serpentine. The grain is at rotation 2 where its highly inclined extinction can be checked.

Figure 2.40 Specimen of tremolite. *Wikimedia Commons photograph by Didier Descouens.*

Figure 2.41 Carved and polished green serpentine marble from Connemara, western Ireland, decorating the interior of the Museum Building at Trinity College Dublin.

transitional composition between limestone and shale. Such a sedimentary rock is known as **marl**. When marl is metamorphosed the resulting marble will contain silicates with calcium (from calcite) and aluminium (from shale), such as the minerals epidote, grossular (calcium garnet), or perhaps anorthite (Ca-plagioclase). **Epidote** is a dense yellowish green mineral with the chemical composition $Ca_2(Al,Fe^{3+})Al_2(SiO_4)_3(OH)$, as outlined in section A2.4.1 in Appendix 2. The iron it contains is Fe^{3+}, and it substitutes for one of the three Al^{3+} ions in the formula, but little Fe^{3+}occurs in marbles. The appearance of epidote in thin section will be described more fully in the forthcoming section on metamorphosed basic igneous rocks. **Grossular** garnet has a somewhat related formula to epidote, $Ca_3Al_2(SiO_4)_3$, and

anorthite is the Ca end-member of plagioclase with the formula $CaAl_2Si_2O_8$ (see Appendix 2). These two minerals occur particularly in impure marbles in contact aureoles, and more will be said about them in chapter 4.

2.3.3 Metasediments with mixed compositions

Sedimentary rocks with transitional compositions exist not only between limestone and quartz sandstone, and between limestone and shale, as just described, but they also exist between quartz sandstone and shale where intermediate rocks include ordinary sandstone, muddy sandstone (called greywacke), and siltstone. Compositions in the middle of this range give rise to semipelitic metamorphic rocks.

Where the amount of 'contaminating' mud or quartz in limestone is considerable, metamorphic reactions may consume all the calcite and dolomite, leaving a rock made entirely from silicates. Such a rock is known simply as a **calc-silicate** rock, and, depending on its protolith, it may be dominated by Ca-Al silicates like epidote, grossular and anorthite, or by Ca-Mg silicates like forsterite, diopside and tremolite.

The full variation in the composition of the three common kinds of sedimentary rock can be represented on a **triangular diagram** with sand, carbonate and mud at its three corners (Fig. 2.42). Using this triangular diagram, the names of sedimentary rocks are shown in Figure 2.43 and the names of the

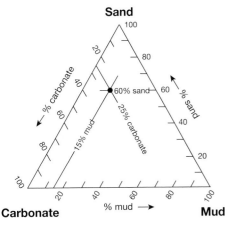

Figure 2.42 (above) Triangular diagram with the three main components of sedimentary rocks – sand, carbonate and mud – at its corners. Any point within the triangle represents some combination of the three corner components, and the sum of the three will always be 100%. A sedimentary rock with 60% sand, 25% carbonate and 15% mud is shown as an example.

Figure 2.43 (right) The same triangle as in Figure 2.42 showing, on the left, the names of sedimentary rocks with different combinations of sand, carbonate and mud, and, on the right, the names of the equivalent metasediments.

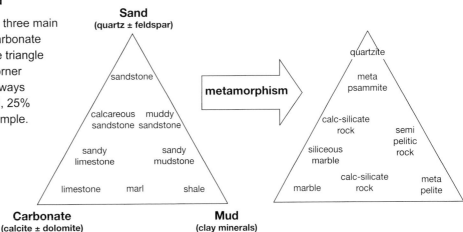

equivalent metamorphic rocks are shown on a corresponding triangle beside it.

Before leaving metasediments, it must be acknowledged that shale, sandstone and limestone are not the only kinds of sedimentary rock. They are certainly the most abundant kinds, but other less common kinds exist, such as ironstone, coal, rock salt and gypsum. Space does not permit the detailed description here of the metamorphic products of these less common sedimentary rocks. Meta-ironstones are known that contain iron oxide (e.g. magnetite), and iron-rich silicates such as garnet, and iron olivine. Diagenetic changes to coal are well known, where bituminous coal changes into hard anthracite. With low-grade metamorphism, anthracite would change into crystalline carbon in the form of graphite rock. Gypsum loses its water of crystallization and becomes anhydrite rock ($CaSO_4$), but rock salt does not change.

2.4 Metabasite

Metabasite is a name coined to include all metamorphic rocks with a **basic igneous** protolith, i.e. rocks with between 45% and 52% SiO_2 by weight, regardless of whether the protolith was basalt (fine-grained frozen lava), **dolerite** (medium-grained frozen magma in small intrusions like sills and dykes), or **gabbro** (coarse-grained frozen magma in large intrusions), or even basic volcanic ash. Basic (basaltic) magma is produced in copious volumes when the mantle undergoes decompression melting, as explained in section A1.2.1 of Appendix 1. Basic igneous rocks are the most abundant kind of igneous rocks on the planet, comprising almost the entire oceanic crust as well as much of the continental crust. Metabasites may preserve features from the protolith such as igneous textures, the sheet-like geometry of **sills** and **dykes**, and the ovoid form of pillows in **pillow basalt** (Fig. 2.44).

Basic igneous rocks contain two essential minerals, **plagioclase** and **augite** (pronounced **or**-gite, usually with a hard 'g'). Augite is an example of **clinopyroxene** (see Appendix 2). These minerals can be seen in the photograph of gabbro (Fig. 2.45), where the plagioclase forms pale grey tablet-shaped crystals, while the augite occurs as dark blocky grains lodged between them. The rock as a whole is dark-coloured. Both minerals may break along cleavage planes and so the grains in a hand specimen of basalt or gabbro glitter slightly as the specimen is turned and the cleaved surfaces catch the light.

Figure 2.44 Metamorphosed pillow basalt from the Neoarchaean Tisdale Formation, Abitibi Greenstone Belt, near Timmins, Ontario, Canada. *Photo courtesy of Balz Kamber.*

Figure 2.45 A specimen of gabbro, composed of calcic plagioclase (pale elongate outlines) and augite (dark).

Plagioclase and augite in a thin section of dolerite can be seen in Figure 3.10.

The plagioclase in basic igneous rocks is calcic plagioclase; it has more than 50% of the calcium end-member, anorthite, in the solid solution series between **albite** ($NaAlSi_3O_8$) and anorthite ($CaAl_2Si_2O_8$) as explained in Appendix 2 (sections A2.2 and A2.4.5). Augite is a variety of clinopyroxene with the formula, $Ca(Mg,Fe)Si_2O_6$, modified by limited atomic substitution of Ca by Na, of Ca by Mg, of Mg by Al and of Si by Al, again as outlined in Appendix 2 (section A2.4.2). Therefore, basic igneous rocks contain seven of the eight main elements of the continental crust (all except K), and metabasites have a correspondingly large variety of mineral assemblages, six of which are now described.

2.4.1 Six kinds of metabasite from regional metamorphic belts

The six different kinds of metabasite described here are all found in regional metamorphic belts, and each has its own name. They are called greenschist, epidote amphibolite, amphibolite, pyroxene granulite, blueschist and eclogite (pronounced **ek**-log-ite). The minerals in them come from a list of nine separate minerals. Five have been met already – chlorite, epidote, plagioclase, garnet and clinopyroxene. The four new ones are orthopyroxene, and three kinds of clinoamphibole called actinolite, hornblende, and glaucophane (see Appendix 2). Each metabasite contains at least one strongly coloured Fe-Mg-bearing mineral, so they are all coloured rocks, from green to almost black.

Greenschist

Greenschist, as its name implies, is a green metabasite with a schistose texture. It is a low-grade rock, found in the chlorite and biotite zones, and is usually fine-grained. Its colour is due mostly to an abundance of the green sheet silicate mineral chlorite, $(Mg,Fe)_5AlSi_4AlO_{10}(OH)_8$, which was met earlier in slate and schist. Greenschist also contains the yellow-green mineral epidote, $Ca_2(Al,Fe^{3+})Al_2(SiO_4)_3(OH)$, which was just met in metamorphosed marls. It always contains albite (the Na-rich end-member of plagioclase; see Appendix 2), formula $NaAlSi_3O_8$, and it may contain the green mineral **actinolite**.

Actinolite has not been met before. It forms long green prisms and is one of the clinoamphiboles described in Appendix 2. Its formula is $Ca_2(Mg,Fe)_5Si_8O_{22}(OH)_2$, so it is almost the same as tremolite (which was just met in siliceous dolomitic marbles) but with substitution of Fe for Mg.

In some cases, rocks called greenschist are not actually schist. They are non-foliated, so the name *greenstone* is really more appropriate. **Greenstone** can result from hydrothermal

metamorphism at ocean ridges (see section 1.3.2 of chapter 1), where there is no directed stress. Here, basalt or dolerite can be completely replaced by the greenschist minerals (albite, epidote, chlorite ± actinolite) without losing the original igneous texture. The name **spilite** was used historically for such rocks. In the nineteenth and earlier twentieth centuries, it was wrongly believed that spilite crystallized directly from a water-rich basaltic magma, because the igneous texture is so faithfully replicated.

Greenschist in thin section is shown in Figure 2.46 (VM GeoLab M15). It comes from a metamorphosed dolerite sill close to the NW margin of the Grampian orogenic belt in Ireland. It is dominated by porphyroblasts (large grains) of chlorite, which is pleochroic from green to almost colourless in PPL, and, in XP, has dark, murky purplish interference colours and parallel extinction. Other minerals in the rock are fine-grained, and occur in thin parallel bands. Clusters

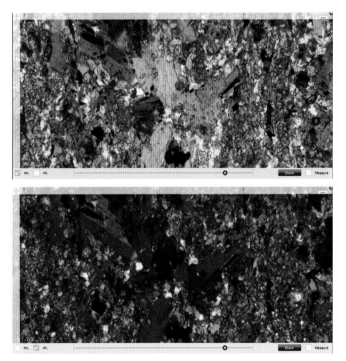

Figure 2.46 Screen shots of VM GeoLab, M15 – greenschist in PPL (top) and XP. Large grains that are green in PPL and deep, murky purple to almost black in XP are chlorite. Bright colours in XP are tiny granules of epidote. Width of field is 1mm.

of granular epidote (high relief in PPL and bright colours in XP) and streaks of tiny colourless grains with low relief, which probably include both quartz and albite, are visible. Some opaque grains are also present. Actinolite, a mineral that commonly occurs in greenschist, is not present here. The texture of this rock at low magnification (seen online) suggests that the original dolerite first became intensely flattened and streaked out, and that the chlorite porphyroblasts grew afterwards.

The pleochroism of chlorite can be examined online at rotation 1, where a second pleochroic mineral, biotite, turns from pale yellow to almost black, and it has layers of chlorite sandwiched within it. Biotite contains potassium so it is not normally prominent in metabasites; potassium is a rare element in basalt. Its presence here hints that some potassium may have been introduced to the rock during metamorphism. The process of adding new elements like this was noted briefly in chapter 1. It is called **metasomatism**.

Epidote amphibolite

Epidote amphibolite occurs at medium metamorphic grade, in the garnet zone. Like greenschist, it contains epidote and albite. However, instead of chlorite it contains a different hydrous mineral, **hornblende**.

Hornblende is the most abundant variety of **clinoamphibole**. Its chemistry is like that of actinolite, but with Al substituting for both Mg and Si (see Appendix 2) giving a formula such as $Ca_2(Fe,Mg)_4Al(Si_7Al)O_{22}(OH)_2$. It can also take in Na and K, making it one of very few minerals that can contain all eight of the main elements of the continental crust. Its crystals are black prisms (Fig. 2.47) with a diamond-shaped cross-section and two cleavage directions that intersect at about 60°.

A thin section of epidote amphibolite is shown in Figure 2.48. It is from the same rock as VM GeoLab, M02, which was collected from within the garnet zone of the Grampian orogenic belt in County Mayo, Ireland, not far from the garnet-mica-schist, M08. The hornblende, epidote and albite are distinctive. While hornblende is black as a mineral specimen (Fig. 2.47) in PPL it is green and pleochroic. In XP it has slightly (10° to 20°) inclined extinction and has interference colours up to first-order orange. (It can be distinguished from chlorite, seen in the previous rock, since the latter has parallel extinction and dark purple-grey interference colours). Epidote stands out in XP with bright second-order interference colours. In PPL it

Figure 2.47 A prismatic fragment of hornblende, broken along cleavage planes from a larger crystal.

Figure 2.48 Epidote amphibolite, from the same rock as GeoLab, M02, in PPL (top) and XP. Hornblende is green in PPL. Epidote granules have bright interference colours in XP. Albite forms a single irregularly shaped grain, uniformly grey in XP. Titanite forms the triangular grain with very high relief and is high-order white in XP. Width of field is about 1mm.

has high relief, is colourless to pale yellow, and has no cleavage. It resembles the mineral olivine but, being hydrous, it is often accompanied by other hydrous minerals such as green chlorite or hornblende, which olivine, being anhydrous, rarely is. The albite in PPL has low relief and is colourless, while in XP it is uniformly grey with unusual outlines that protrude between, and seem to have partly replaced, the surrounding hornblende and epidote. It also has inclusions of these last two minerals. The conspicuous pale brownish triangular grain with high relief, at top centre, half enclosed by plagioclase in Figure 2.48, is **titanite**, a titanium (Ti) bearing accessory mineral whose formula is $CaTiSiO_5$. It is distinctive because in XP it has high-order white interference colours, like those of calcite, that do not change when a sensitive tint plate is inserted into the light path.

Elsewhere in the rock, but not visible in Figure 2.48, is a little quartz. Seen online quartz looks similar to albite, but commonly shows strain shadowing (check rotation 2, right side) and has no inclusions. Also online, at low magnification, the rock can be seen to be foliated with sub-parallel hornblende grains, and with layers of epidote alternating with layers of hornblende.

Amphibolite

Amphibolite is a medium-grade metabasite associated in the field with rocks of the staurolite, kyanite and sillimanite zones. It is dark green to black in colour, sometimes speckled, and it is composed of just two main minerals – hornblende, which is dark green or black, and intermediate plagioclase, which is white or grey (Fig. 2.49). Other minerals that may be present in small amounts include titanite, garnet, quartz, biotite and iron oxides.

A thin section of amphibolite, different from that in Figure 2.49, is shown in Figure 2.50. It contains garnet, and comes from the Grampian belt near Glenelg in NW Scotland. Hornblende is pleochroic; its colour varies, grain to grain, from greeny brown to pale yellow. Many of the grains are prismatic and end-on, displaying the diamond-shaped cross-section and the characteristic two cleavage directions intersecting at roughly 60°. The abundance of end-on views shows that the hornblende prisms are aligned with each other in the rock, with a texture known as a **mineral lineation**. The colourless mineral is intermediate plagioclase. Garnet is the large, rounded grain with high relief, top centre. A flake of brown

Figure 2.49 A specimen of amphibolite from a metamorphosed dolerite dyke in NW Scotland. White grains are of plagioclase and were formerly phenocrysts (large crystals) in the dolerite.

Figure 2.50 Thin section of amphibolite with garnet in PPL. Field of view is about 1mm wide. For a description see the text.

biotite is present on the right, and the tiny granules with high relief on the left are titanite.

Pyroxene granulite

Pyroxene granulite is a high-grade metabasite. It is a coarse-grained, dark-coloured granular rock consisting of the three minerals – augite (clinopyroxene), orthopyroxene and plagioclase. **Orthopyroxene** has not been met before. It is a brown mineral with the formula $(Mg,Fe)SiO_3$. It differs from clinopyroxene ($Ca(Mg.Fe)Si_2O_6$) in containing no calcium, as described in section A2.4.2 of Appendix 2. Like all pyroxenes it has two cleavage directions that intersect at about 90°.

An example of pyroxene granulite in thin section (VM GeoLab M24) is shown in Figure 2.51. It is from the Lewisian gneiss, an ancient high-grade rock unit west of the Grampian belt in NW Scotland. In PPL the clear, colourless mineral with low relief is plagioclase, which shows distinctive lamellar twinning (parallel grey stripes) in XP. The two kinds of pyroxene can be distinguished in PPL because the orthopyroxene is weakly pleochroic from pale pink to pale green, whereas the augite is pale greyish green and not pleochroic. The pleochroism should be checked online at rotation 2. The two pyroxenes can also be distinguished online in XP because orthopyroxene has parallel extinction and first-order grey

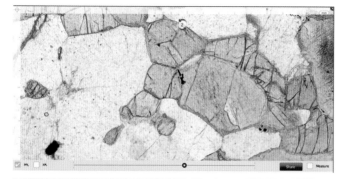

Figure 2.51 Screen shots in PPL (upper image) and XP of VM GeoLab Two pyroxene granulite, M24, from the Lewisian gneiss in NW Scotland. Minerals are plagioclase, clinopyroxene and orthopyroxene, and the texture is granoblastic. For a full description see the text. The field is 2mm wide.

interference colours, whereas augite, like other kinds of clinopyroxene, has highly inclined extinction, and colours up to second-order blue (as at rotation 1).

All the grains have polygonal outlines and are roughly the same size (Fig. 2.51). This texture is described as **granoblastic**, and is seen in various metamorphic rocks. Basalt and gabbro have similar minerals to pyroxene granulite (pyroxene and calcic plagioclase) but they have an igneous texture with randomly orientated, elongate plagioclase shapes.

In Figure 2.51 green fringes around some of the pyroxene margins are where it has been altered to hornblende. Hornblende is a hydrous mineral, so its formation here reflects the ingress of a little water into what is otherwise an anhydrous rock. This is another example of retrograde metamorphism, similar to the one described in the forsterite marble where serpentine was produced along cracks in the forsterite grains.

Blueschist

Blueschist is a metabasite composed, commonly, of two minerals, glaucophane and epidote, and it usually has a schistose texture (Fig. 2.52).

Glaucophane (pronounced **glor**-ka-fane) is another new mineral here. It is a deep blue member of the clinoamphibole group of minerals, and contains a lot of sodium and aluminium. Its formula can be obtained by substituting Na for Ca, and Al for Mg in the formula for actinolite, giving $Na_2Al_2(Mg,Fe)_3Si_8O_{22}(OH)_2$.

A thin section of blueschist collected from the same locality as the hand specimen in Fig. 2.52 is shown in Figure 2.53. The two minerals, glaucophane and epidote, are easy to distinguish. Glaucophane is quite striking in PPL, appearing blue, purple and colourless. Its pleochroism can be examined online at rotation 3. The area around rotation 3, shown in Figure 2.53, is largely of glaucophane, and several grains show the 60° cleavage intersection characteristic of amphiboles. In XP the glaucophane shows bright interference colours, but this is not normal; glaucophane should show grey to creamy white first-order interference colours. The bright colours are seen here because the thin section is much thicker than the standard 30 microns.

Figure 2.52 A specimen of epidote-rich blueschist, about 10cm wide, from the Ile de Groix, Brittany, France. It is banded, and the bands have been crumpled into folds. The dark blue-grey bands are glaucophane and the yellowish-green bands are epidote.

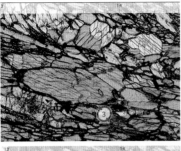

Figure 2.53 Screen shots of VM GeoLab Glaucophane schist, M12 from Ile de Groix in PPL (top) and XP. It is largely glaucophane, with a large grain of epidote at the lower left. Field is 1.5mm wide.

Figure 2.54 A specimen of eclogite from near the village of Almklovdal, Nordfjord in western Norway. It consists of red garnet and green clinopyroxene. The specimen is 10cm wide.

An elongated grain of epidote lies near the lower left corner. In XP its bright second-order interference colours are very different between the grain's core and its rim. This spatial variation is common in epidote, and is because the birefringence (see Appendix 3) of epidote increases enormously for just a small amount of substitution of iron (as Fe^{3+}) for aluminium (Al^{3+}) in the epidote formula, $Ca_2(Al,Fe^{3+})_3(SiO_4)_3(OH)$.

Glaucophane schist differs in an important way from the four metabasites described before it: it has no plagioclase. The Na and Al otherwise found in plagioclase have gone into the glaucophane instead, and the Ca and Al from plagioclase are in epidote.

Eclogite

Eclogite is a dense and brightly coloured metabasite, consisting of red garnet and green clinopyroxene (Fig. 2.54). It is usually coarse-grained without schistosity. The garnet has some Ca and Mg substituting for Fe, giving it the general formula $(Fe,Mg,Ca)_3Al_2(SiO_4)_3$. The clinopyroxene contains a considerable amount of sodium and aluminium, and has the odd-sounding name **omphacite** (pronounced **om**-fa-site). Its general formula is $(Na,Ca)(Al,Mg,Fe)Si_2O_6$.

A thin section of eclogite can be examined online as VM GeoLab M26, and a view of this rock is shown in Figure 2.55. Here, the garnet forms rounded grains that are pale pink. Its

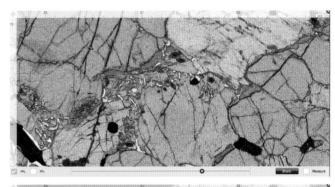

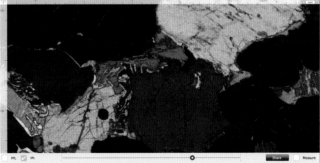

Figure 2.55 Screen shots in PPL (top) and XP of eclogite from the Glenelg district of NW Scotland (VM GeoLab M26). The width is 3mm. It is described in the text.

36

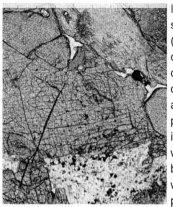

Figure 2.56 Eclogite – M26 showing detail of omphacite (Na-Al-rich clinopyroxene) cracked into small squares owing to two cleavage directions intersecting at about 90°. The lower part of the omphacite was inadvertently plucked out while the thin section was being made leaving a 'hole' with bits of black grinding paste. Width of field is 1mm.

high relief, lack of cleavage and isotropic nature make it easy to identify. The omphacite is faintly green in PPL. Like other kinds of clinopyroxene, it may show parallel cleavage cracks, and in XP it shows grey, cream, orange, purple and blue interference colours, and has highly inclined extinction to the cleavage (see rotation 1). One grain near the lower edge of the thin section shows the intersection of two cleavage directions at about 90°, characteristic of pyroxene (Fig. 2.56).

Also present are some quartz and rutile. The quartz shows unmistakable **strain shadowing** at rotation 2. **Rutile** is titanium dioxide, formula TiO_2. It is an accessory mineral that is deep brown with very high relief. It forms sporadic, small rounded grains and is normally found in eclogite.

Eclogite, like glaucophane schist, contains no plagioclase. However, some plagioclase with green hornblende does occur in M26 along boundaries between garnet and omphacite (Fig 2.55). The two minerals seem to have grown *after* the eclogite had formed. Plagioclase and hornblende together make the rock amphibolite. Since hornblende is hydrous and garnet and omphacite are not, it appears that a little water has gained access to the eclogite and converted it locally, along grain boundaries, to amphibolite. This is another example of retrograde alteration.

2.4.2 The ACF triangle for minerals in metabasites

A triangular diagram has been devised that is suitable for showing the chemical compositions of minerals in metabasites, and in other rocks too. The triangular diagram works the same way as the one for sedimentary rock compositions in Figure 2.42. Since there are eight main elements, and only

three corners to a triangle, the method of plotting compositions involves omitting some elements and combining others. Oxygen and silicon are omitted, and so are the alkali elements Na and K. Iron and magnesium substitute for each other, so they are combined. This means that Al can go at one corner, Ca at another and combined (Fe + Mg) at the third. The triangular diagram is known as an **ACF** plot where A stands for Al, C for Ca, and F for combined Fe + Mg. It shows mineral compositions in terms of the relative amounts of the four elements Al, Ca, and combined Fe and Mg (as numbers of atoms). Rock compositions can also be plotted.

The positions of all nine metabasite minerals are shown in Figure 2.57. Epidote can serve to illustrate how a mineral's composition is plotted. Taking its formula simply as $Ca_2Al_3Si_3O_{12}(OH)$, there are two atoms of Ca and three atoms of Al giving a total of five atoms that 'count' (there are zero Fe and Mg atoms). Since three of the five are aluminium, epidote plots on the edge between Ca and Al, three-fifths or 60% up from the Ca corner towards the Al corner. The formulae of the nine minerals named on Figure 2.57 are listed in Appendix 2 (Table A2.1). Hornblende alone has elements from all three

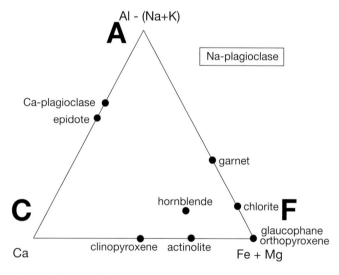

Figure 2.57 An ACF triangular diagram showing the compositions of nine minerals that occur in metabasites (a tenth mineral, Na-plagioclase, is named separately) in terms of the numbers of Al, Ca and Fe+Mg atoms each mineral contains. The top corner is labelled Al – (Na + K) for reasons explained in the text.

corners, and lies inside the triangle; the others plot on the edges or at the corners.

One proviso in plotting a composition concerns minerals with the alkali elements, Na and K. These elements are in Na and K-feldspars, and they can substitute into clinoamphibole and clinopyroxene. In all cases they are balanced by an equal number of Al atoms in the mineral formula. Since Na and K are omitted from the triangular plot, then the convention is that the Al that goes with them is also omitted. So, before calculating the position of a mineral with Na or K in its formula, the number of Al atoms in the formula is first reduced by the combined sum of (Na + K). Thus the A corner of the triangle strictly corresponds to Al − (Na + K). As an example, glaucophane's formula is $Na_2(Mg,Fe)_3Al_2Si_8O_{22}(OH)_2$. It contains 2 Al atoms, but it also contains 2 Na atoms, so the Al here is omitted, and the formula of glaucophane, having no Ca, and with its Al equal to its Na, simply plots at the 'F' corner.

Plagioclase feldspar also needs some explanation. Its atomic substitution is described in Appendix 2. The Ca end-member, $CaAl_2Si_2O_8$, with one Ca and two Al atoms, plots two-thirds the way up the edge between C and A, just above the position of epidote. The Na-end member, however, cannot be shown on the triangle because its formula, $NaAlSi_3O_8$, has Al equal to Na. The convention is to write the name *Na-plagioclase* or *albite* separately, beside the triangle, where albite is present in the rock, and to be aware that where Ca-plagioclase is shown on the triangle, this usually refers to the Ca end-member within intermediate plagioclase.

A final point to note regarding the ACF triangle is that, historically, a different version of it has been widely used. This has oxygen combined with the elements at the corners, giving the oxides, Al_2O_3, CaO and (FeO + MgO) at the three corners. This system means that the positions of Al-bearing minerals are all shifted systematically away from the A corner because they are plotted according to the numbers of Al, Ca and (Fe+Mg) oxide molecules, and each molecule of Al_2O_3 has *two atoms* of Al.

The minerals in each of the six metabasites are plotted in six separate triangles in Figure 2.58, where a green disc marks the fixed metabasite protolith composition. This shows that, despite their diversity of mineral assemblage, all six metabasites have identical chemistry.

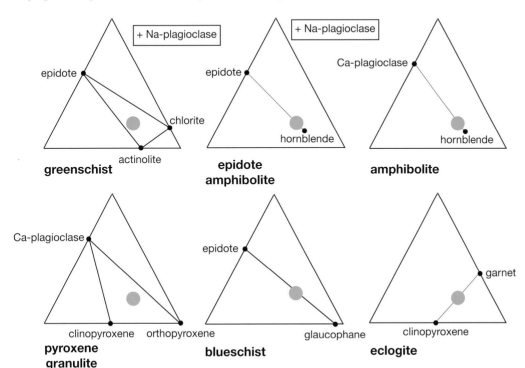

Figure 2.58 The same triangle as in Fig. 2.57 repeated six times, showing the minerals present in each metabasite, and how the protolith composition (filled green circle) in each case falls on the line, or within the triangle, joining the minerals.

2.4.3 P–T stability of metabasites, and metamorphic facies

The four kinds of plagioclase-bearing metabasite, namely greenschist, epidote amphibolite, amphibolite and pyroxene granulite, were introduced above in the order of increasing metamorphic grade. The progressive changes to the minerals in them, and how the changes link with Barrow zones, are summarized in Figure 2.59. This diagram shows how greenschist (with Na-plagioclase, epidote, actinolite and chlorite) spans the chlorite and biotite zones, and is a low-grade rock. It shows that epidote amphibolite (with hornblende having replaced actinolite, and with chlorite in decline, but with Na-plagioclase and epidote still present) corresponds roughly with the garnet zone. It shows that epidote stops where amphibolite proper starts (with hornblende and intermediate Na-Ca-plagioclase). Evidently, the Ca and Al released from epidote join the Na-plagioclase, turning it into intermediate plagioclase. Amphibolite is a medium-grade rock that straddles the staurolite, kyanite and sillimanite zones. Finally, Figure 2.59 shows that moving to high-grade conditions (beyond the sillimanite zone), pyroxene granulite

appears. Here the hornblende fades out with the arrival of orthopyroxene and clinopyroxene. Intermediate plagioclase continues. Taken together, the progressive changes involve loss of water, starting from water-rich greenschist and ending with anhydrous pyroxene granulite. The changes are also, incidentally, accompanied by an increase in grain size.

What of blueschist and eclogite? How do they relate to metamorphic grade? Since neither contains plagioclase, which has a relatively low density, blueschist and eclogite are significantly denser than the other metabasites. High density is favoured by high pressure, so it seems that blueschist and eclogite are made at higher pressures than the other metabasites, and presumably over the same range of temperatures. Also, since blueschist contains hydrous minerals and eclogite does not, then blueschist is taken to form at lower temperatures than eclogite. The relative pressure and temperature (P–T) conditions for all six metabasites are shown in a qualitative way in Figure 2.60.

An attempt to quantify the metabasite fields in Figure 2.60 is shown in Figure 2.61. This diagram is based largely on the

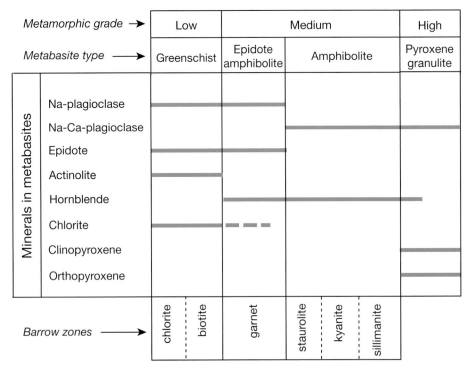

Figure 2.59 Chart showing progressive changes, with increasing grade, to the minerals in the four plagioclase-bearing metabasites. The corresponding Barrow zones (see Fig. 2.28) and terms for metamorphic grade are also shown. The changes are reviewed in the text.

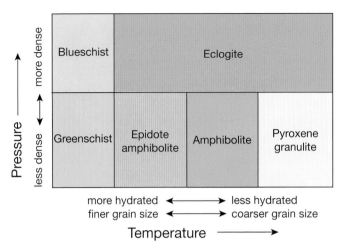

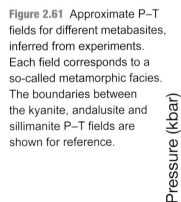

Figure 2.60 Inferred positions of the six metabasites on a pressure–temperature grid.

below about 300°C, so strictly they correspond to the realm of diagenesis rather than metamorphism. They are named '**prehnite-pumpellyite**' and '**zeolite**'. These are not names of metabasites, but names of distinctive minerals that grow in cavities in basic igneous rocks (mainly basalt and volcanic ash) at these low temperatures. Prehnite (pronounced **prey-nite**) and pumpellyite are dense hydrated calcium-aluminium bearing silicates, chemically quite similar to epidote. Zeolite is the name for a group of low-density hydrous framework silicates with Na, Ca and Al. Zeolites contain water as whole water molecules that reside in cage-like micropores within the framework structure.

The fields for the five metabasites (six if epidote amphibolite were to be treated separately) plus the zeolite and prehnite-pumpellyite fields in Figure 2.61 provide the basis for a method of classifying metamorphic rocks according to P–T conditions. This is the **metamorphic facies** classification scheme that was mentioned briefly in chapter 1. The word 'facies' is appended to each of the names in Figure 2.61. Thus one can talk of the blueschist facies, the amphibolite facies, the zeolite facies, and so on. The pyroxene granulite facies usually has its name shortened to just *granulite facies*. This gives seven metamorphic facies, or eight if the epidote amphibolite facies is included.

results of **experimental petrology** – the synthesis of rocks and minerals under controlled P–T conditions in the laboratory – which will be described in chapter 3. In Figure 2.61 the field for epidote amphibolite is not shown; it has been combined with the one labelled amphibolite. Two other fields have been added. Both correspond to low temperatures,

Figure 2.61 Approximate P–T fields for different metabasites, inferred from experiments. Each field corresponds to a so-called metamorphic facies. The boundaries between the kyanite, andalusite and sillimanite P–T fields are shown for reference.

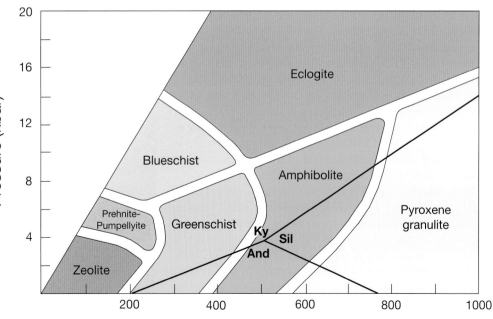

Each facies is a kind of P–T window; it comprises metamorphic rocks, made from many different protoliths, that accompany one of the eight kinds of metabasite, and which were presumably metamorphosed under the same range of P–T conditions as that metabasite. As an example, the pyroxene granulite facies includes metapelites, marbles and other rocks that are found in the same tract of country as pyroxene granulite and which, therefore, are presumed to have formed in the P–T field of pyroxene granulite. The metamorphic facies classification system is not strictly quantitative; it is used, like grades and zones, to give a general sense of the kinds of pressure and temperature at which metamorphism occurred. The zeolite and prehnite-pumpellyite facies, as noted above, correspond to diagenesis and the anchizone in pelitic rocks.

2.4.4 A metabasite made by contact metamorphism

This account of metabasites now changes tack and switches from regional metamorphism to thermal metamorphism. The example chosen for description is an amphibolite (GeoLab M01). It is not from a thermal aureole as such, but comes from a huge **xenolith** (fragment of country rock), perhaps 200 metres across, surrounded by granite at Lough Nagilly in County Donegal, NW Ireland. This xenolith is probably a foundered block from the roof of the magma chamber. A thin section of it is shown in Figure 2.62, where hornblende, intermediate plagioclase and a few opaque grains can be seen. The hornblende in PPL is pleochroic, between deep brown-green, green and straw-coloured (which can be checked online), and in XP it shows the normal first-order into second-order interference colours and inclined extinction. Figure 2.63 shows a rare grain with the diagnostic intersection of two cleavage directions at about 60°. Plagioclase shows lamellar twinning in many grains. Opaque grains are present; they may be made of magnetite, Fe_3O_4. There is absolutely no grain alignment, consistent with thermal metamorphism of a block immersed in granite where directed stress would have been absent.

The thin section appears to show, in the grain distribution, a pattern that has been inherited from the protolith. Elongate, randomly orientated outlines made by grains and clusters of plagioclase can be picked out in PPL, and are presumably a legacy of the plagioclase plates in the dolerite protolith. If so, this is an example of a **relict igneous texture**. In some cases, seen in XP, the original plagioclase twin lamellae appear to have survived.

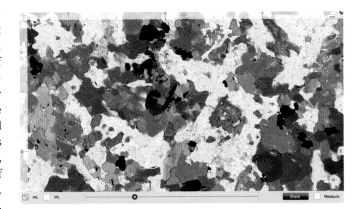

Figure 2.62 Screen shots of VM GeoLab Amphibolite, M01, from Lough Nagilly, County Donegal, in PPL (top image) and XP. Width of field is 2.5mm. The randomly orientated plagioclase laths, best seen in XP, are probably a relict texture, inherited from a dolerite protolith.

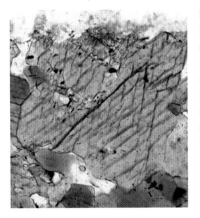

Figure 2.63 Hornblende in M01 on the top edge of the thin section showing the intersection of two sets of cleavage planes about 60° apart, diagnostic of amphiboles such as hornblende. Width of field is 0.5mm.

2.5 Metagranite

Granite is a pale-coloured, coarse-grained igneous rock made of the three minerals – quartz, Na-rich plagioclase (albite) and potassium feldspar, in roughly equal amounts. It usually also contains one or more of the minerals muscovite, biotite and hornblende, in small amounts. It has more than 63% by weight of SiO_2 and, since SiO_2 is the oxide of silicic acid, granite has traditionally been called an **acidic igneous rock.** (For the same reason silica-poor igneous rocks, like basalt, are classified as basic.) A broken piece of granite will tend to sparkle as the light catches the flat cleavage surfaces of mica and feldspar, though quartz breaks unevenly, like glass, and does not sparkle.

Granite occurs almost exclusively in the continental crust, and is believed to be produced when rocks of the deep continental crust, particularly deeply buried shale, become extremely hot and start to melt, as suggested in the field by the layers and blobs of granite in migmatite. The resulting granitic liquid, which has a low density, is thought to gather into underground pools and then migrate upwards. It typically intrudes the upper crust as plutons 10km or more across (e.g. Fig. 1.8), or breaks the surface as volcanic domes of **rhyolite.**

Figure 2.64 Sawn surface of a specimen of granitic gneiss from Ardgour in NW Scotland. The granite was intruded before the Grampian orogeny, and was deformed by that orogeny. Its aligned black streaks contain biotite and hornblende. Width of specimen is 8cm.

2.5.1 Granitic gneiss and orthogneiss

When granite becomes metamorphosed its minerals change very little, which suggests that they are already stable under most metamorphic conditions. This is not surprising, considering that granitic magma cools slowly, and it probably continues to **crystallize** and harden down to temperatures in the region of 600°C to 700°C when the last little bit of melt finally solidifies. These temperatures are those of amphibolite facies metamorphism.

However, while the minerals stay little changed, the texture of granite is often modified by directed stress. With moderate stress, either during the late stages of magma solidification or during subsequent metamorphism, grains of quartz and feldspar may become flattened, and plates of muscovite and biotite, and prisms of hornblende may end up parallel to each other, giving the rock a foliated texture. The resulting rock is granitic gneiss (Fig. 2.64).

Granitic gneiss is an example of **orthogneiss.** Orthogneiss is a term for gneiss derived from any igneous protolith, be it granitic or basaltic or any other composition. (The complementary term for gneiss with a sedimentary protolith is **paragneiss.**) As it happens, most of the orthogneiss in the continental crust is neither granitic nor basaltic, but has an intermediate composition. **Intermediate igneous rocks** are defined as those having between 52% and 63% of silica by weight, and include the coarse-grained rock called **diorite**, and its fine-grained volcanic equivalent called **andesite**. When these are metamorphosed they turn into intermediate gneiss (dioritic gneiss), just as granite turns into granitic gneiss. As with granite, the minerals in diorite tend not to change greatly during metamorphism. A medium-grade dioritic gneiss consists of intermediate plagioclase and hornblende, much the same as diorite; it is rather like plagioclase-rich amphibolite. At high grade the hornblende is replaced by pyroxene, just as it is in high-grade metabasites when amphibolite changes to pyroxene granulite.

2.5.2 Dynamic metamorphism of granite

Dynamic metamorphism was introduced briefly in chapter 1 (section 1.3). It is caused by movement within shear zones, which are deep-seated faults, and it leads to intense deformation of rocks, and a general reduction in their grain size. It affects all kinds of rock, but granite is a suitable protolith to describe its effects.

Figure 2.65 Lineated texture in paving slabs of granitic gneiss in a temple courtyard at Bumthang, Kingdom of Bhutan. The texture resulted from flattening of quartz and feldspar grains parallel to the plane of the slabs, and enormous stretching of the grains parallel to their length. Width of field is 70cm.

Figure 2.66 Granitic augen gneiss from near Emersson Lake in the Swiss Alps near Chamonix.

During dynamic metamorphism, granite may become stretched and flattened at the same time, resulting in a lineation and a foliation together in the same rock. The lineation in this case is sometimes called a **stretching lineation**. An example from the Himalayas in the Kingdom of Bhutan is shown in Figure 2.65. Here, the long white streaks are made of quartz and feldspar. Large grains of these two minerals were smeared out as the Indian tectonic plate was being forced inexorably beneath the Asian plate to cause thickening of the continental crust and the formation of the Himalayas. The direction of the lineation, seen in rock outcrop, points consistently downwards to the north, parallel to the inclined direction of subduction of the Indian plate.

Another product of the dynamic metamorphism of granite is **augen gneiss**. Here some of the original feldspar and quartz survives as lens-shaped lumps, or augen, with the rest of the rock being finely-banded and weaving between them. A sample of granitic augen gneiss from the Alps is shown in Figure 2.66.

With even more intense dynamic metamorphism, granite will be converted to granitic mylonite. Mylonite is an extremely fine-grained rock that may be thinly banded. Not only granite, but many other kinds of rock, including high-grade gneiss and metapsammite, can be converted to mylonite.

An example of mylonite in thin section is shown in Figure 2.67. This is rock 11 from the VM Open University S339 collection. It is a quartz-feldspar mylonite, probably derived from a feldspathic psammite protolith rather than granite. It comes from close to the Moine thrust on the NW margin of the Grampian orogenic belt in Scotland, not far from where the photograph in Figure 1.12 was taken. Large, elongated and aligned grains of quartz and feldspar have been intensely deformed, and they are enveloped in an extremely fine-grained matrix. The large grains are called **porphyroclasts** (clast is a term meaning *broken fragment*). Unlike porphyroblasts, porphyro*clasts* are large because they began as large grains, and have somehow survived the comminution that affected grains around them. The porphyroclast left of the centre of Figure 2.67 is damaged quartz. It shows an extreme degree of strain shadowing. Porphyroclasts of most minerals are similarly damaged and go into extinction in a piecemeal fashion. For example, severely damaged, strain-shadowed clinopyroxene porphyroclasts are shown later in Figure 3.24. In all these cases, intense strain has somehow turned an original large grain into a number of sub-grains that are not quite aligned, and go into extinction separately when the microscope stage is rotated by a few degrees.

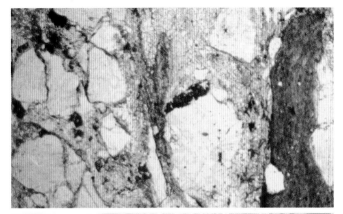

Figure 2.67 Screen shot in XP of VM Open University S339-11, quartz-feldspar mylonite. A large porphyroclast of strain-shadowed quartz is left of centre. Field width 8mm.

Figure 2.68 Quartz-feldspar mylonite in PPL (top) and XP with porphyroclasts of tartan-twinned K-feldspar, and, on the right, a brown strip of pseudotachylite (black in XP). Width of field is about 2mm.

In another example of mylonite from just east of the Moine thrust, shown in Figure 2.68, porphyroclasts of K-feldspar are visible (recognized from the characteristic tartan twinning in XP). Also present, on the right side, is a strip of **pseudotachylite** (pronounced syoo-doh-**taki**-lite). This is rock powdered so finely that it is virtually the same as glass. Indeed, it may actually be glass, formed by frictional melting during sudden fault movement. It is completely black in XP, while in PPL it is brown.

Some rocks that look like mylonite in the field are found in thin section to have strained porphyroclasts, but a matrix between them that is not very fine-grained. Instead, the grains between the porphyroclasts have recrystallized. Recrystallized mylonite like this is called **blastomylonite**. Many examples of augen gneiss are probably blastomylonite, and a hand specimen of one is shown in Figure 2.69. This rock is not a metagranite, but a high-grade garnet-bearing paragneiss (a

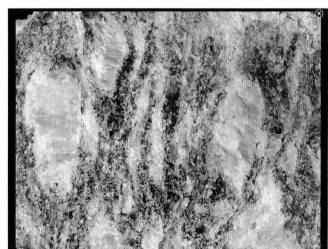

Figure 2.69 Front view and enlarged back view of a small polished slab of augen gneiss collected from Rio de Janeiro, Brazil, by Charles Darwin who described it as 'a true gneiss; a most beautiful rock'. The image is from the collection of Charles Darwin's rocks at virtualmicroscope.org.

metasediment). It was collected by Charles Darwin while on the Voyage of the Beagle, and can be examined in the VM Darwin collection. In the thin section (not shown here, but worth going online to see) a recrystallized streaky matrix of quartz, feldspar and biotite, thought once to have been mylonitic, wends its way between the augen. Much of the quartz occurs as thin sheets, parallel to the foliation, that go into extinction as single, very elongated grains. They are called **quartz ribbons**. A similar example of high-grade pelitic blastomylonite will be described in chapter 5. It comes from near Sligo, in Ireland, and contains kyanite as well as garnet.

2.6 Metaperidotite
2.6.1 Peridotite as a protolith
Peridotite, as was explained in chapter 1 and Appendix 1, is the rock from which the Earth's mantle is made. It is a dense, green rock that is dominated by the mineral olivine, formula $(Mg,Fe)_2SiO_4$. It gets its name from the word *peridote*, which is gem-quality olivine.

Several kinds of peridotite exist. In all of them the olivine is quite close in composition to the magnesium end-member, forsterite, Mg_2SiO_4 which has been seen in dolomitic marble. The most common kind of peridotite, exemplified by the nodule from basalt in Figure A1.3, is named **harzburgite** (pronounced **harts**-burg-ite). It contains some orthopyroxene (which forms the dark grains in Fig. A1.3). Orthopyroxene, like the olivine, is close in composition to its Mg-rich end-member ($MgSiO_3$), named **enstatite**. Harzburgite is a so-called melt residue; it is the kind of peridotite left behind in the mantle after 'normal' peridotite has become partially molten and lost basaltic magma, as explained in Appendix 1. 'Normal' peridotite, prior to partial melting, contains clinopyroxene as well as orthopyroxene, and is called **lherzolite** (pronounced **lurtz**-a-lite). Peridotite with almost nothing but olivine is called **dunite** (pronounced **dunn**-ite).

Some peridotite is not from the mantle but occurs in the lower levels of large intrusions of gabbro where, during the early stages of cooling, crystals of olivine, growing in the magma, sink and accumulate on the floor of the magma chamber. Rocks formed in this way are called **cumulates** (pronounced as in the word *accumulates*).

Peridotite does not fit comfortably into the standard system of rock classification. It is usually grouped with igneous rocks because olivine-rich cumulates form in magma chambers.

Since it has less than 45% by weight of silica (the lower limit for basic igneous rocks), it is classified as an **ultrabasic** igneous rock. But mantle peridotite is not a cumulate and was probably never molten, so how can it be igneous? It would need to be raised to 1800°C at a shallow depth to melt completely, and such conditions do not exist on the planet. Mantle peridotite was probably *always* in the mantle, and merely modified over time by periods of deformation, grain growth and partial melting. It is therefore more logical to regard it as a metamorphic rock whose protolith was the primitive (original) **chondrite**-like peridotite from which the Earth was made at the start of the solar system (see Appendix 1).

While peridotite is essentially a mantle rock, a small amount of it exists in the continental crust and, apart from the peridotite cumulates mentioned above, it was presumably transferred there from the mantle during orogenic processes, in some cases as pieces of so-called ophiolite (see Appendix 1). Its metamorphic products are unusual and quite distinctive. Three kinds of product are described here – anhydrous metaperidotite, hydrous metaperidotite, and carbonate-bearing metaperidotite.

2.6.2 Anhydrous metaperidotite
Mantle peridotite
A thin section of peridotite that travelled directly from the mantle as a nodule in basalt (like the one in the photograph in Fig. A1.3) is shown in Figure 2.70. It was collected in SE Spain. This rock consists mostly of olivine, which is colourless with high relief and no cleavage, and in XP shows interference colours up to mid second order. The two brown grains (in PPL) are isotropic in XP. They are **chrome spinel**, an accessory mineral that is magnesium-aluminium oxide, formula $MgAl_2O_4$, with some of the minor element chromium (Cr^{3+}) substituting for aluminium (Al^{3+}). The chrome spinel is rimmed by plagioclase (a colourless, low-relief mineral in which grey twin lamellae are just about visible in XP). Since chrome spinel, which is dense, appears to be changing on its margins to plagioclase, which is far less dense, it is possible that the change is a response to a drop in pressure while the rock was still in the mantle (before it was carried rapidly to the surface). The large grain in the top left of the field has cleavage and a high extinction angle (which can be seen at rotation 1), so is taken to be clinopyroxene. Though it is not easy to demonstrate, the rock also contains orthopyroxene.

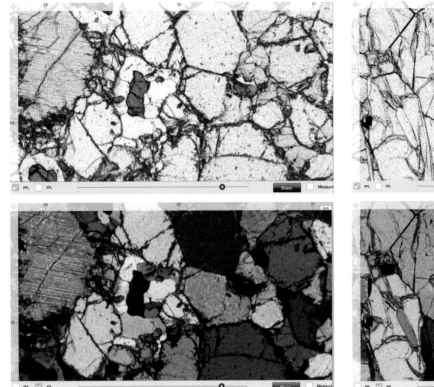

Figure 2.70 Screen shots of VM GeoLab Spinel lherzolite, M19, in PPL (top) and XP. The field is 2.3mm wide. The right half is granoblastic olivine. The brown grain is chrome spinel (an accessory mineral). It is rimmed by plagioclase (colourless in PPL). On the left is a large grain of clinopyroxene.

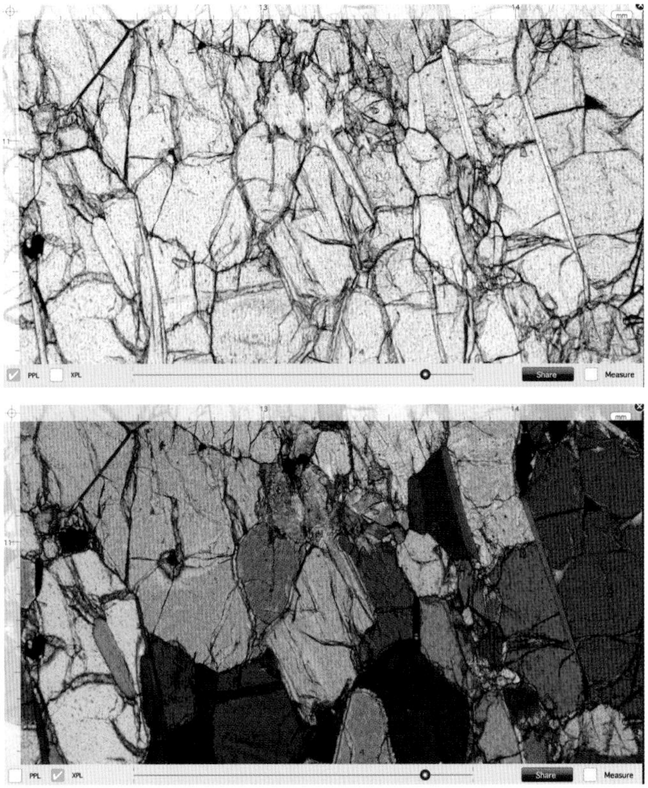

Figure 2.71 Screen shots of GeoLab Dunite, M25, in PPL (top) and XP. It contains aligned grains of olivine, with a little colourless chlorite, talc and orthopyroxene; their distinguishing features are described in the text. Width of field is 2.5mm.

Peridotite with two pyroxenes and spinel is named spinel lherzolite.

Dunite from an orogenic belt

The VM GeoLab thin section, M25, dunite (Fig. 2.71) was collected from a quarry in a kilometre-scale body of dunite in a metamorphic belt in western Norway. The olivine grains here have elongated polygonal outlines and are aligned, consistent with metamorphism under the influence of directed stress. Sharing the alignment are grains of colourless chlorite with long, skinny outlines and low relief. They are grey in XP and have parallel extinction. Chlorite is normally green, but this rock (like nearly all examples of peridotite) is poor

in iron, and it is iron that gives chlorite its colour. One of the long, skinny grains, almost in the centre of the field, is not chlorite; it is bright pink in XP. This is thought to be the magnesium-rich sheet silicate, **talc**. Talc is introduced later as the main mineral in soapstone. The grain on the top edge with grey interference colours is not olivine because it has cleavage. This is probably orthopyroxene. A better example of orthopyroxene can be seen online, in the centre of rotation 1, where its grey interference colour and its parallel extinction support its identification.

Incidentally, the quarry where this rock was collected produces two million tons of crushed dunite per year as a refractory sand for casting molten metal. It was set up many years ago, purportedly on the advice of the pioneering Norwegian geochemist, Victor M. Goldschmidt, who, knowing the extremely high melting point of Mg-rich olivine, recognized its potential for this purpose.

2.6.3 Hydrous metaperidotite
Serpentinite

Large and small pieces of peridotite can become transferred from the mantle to the continental crust during orogenesis, for example as the slabs of ophiolite mentioned in Appendix 1. If water is available, the peridotite readily becomes hydrated under low-grade conditions to give the rock **serpentinite** (Fig. 2.72). Serpentinite is a low-density green to black, waxy-looking rock composed almost entirely of the mineral serpentine. Note that serpentine is the name of the mineral, and '*ite*' is added for the name of the rock.

Serpentine was briefly introduced as a mineral in siliceous dolomitic marble. It is a hydrous magnesian sheet silicate with the formula $Mg_6Si_4O_{10}(OH)_8$. This formula could be halved, and written $Mg_3Si_2O_5(OH)_4$, but the full formula is preferred because the 'Si_4O_{10}' part of it is a reminder that serpentine is a sheet silicate; all sheet silicates have Si_4O_{10} (or Si_3AlO_{10}) in their mineral formulae (see Appendix 2).

Serpentinite has a much lower density than peridotite, so if water gains access to the mantle and converts a large volume of it to serpentinite, the latter becomes buoyant and pushes its way upwards, *en masse*, to shallower levels. As an example,

the Troodos Mountains in Cyprus are made from a huge body of serpentinite, tens of kilometres across. It is part of a giant slab of ophiolite that was formerly sub-seafloor mantle.

In thin section, serpentinite has a distinctive appearance. It can be seen in GeoLab M18 (Fig. 2.73), which comes from the west of Ireland. Its geological setting is described in chapter 5. It consists almost entirely of serpentine, some colourless, some very pale green. Dark brown rounded grains are chrome spinel. The colourless oval patch near rotation 2 is possibly a pseudomorph after an augen-like grain of orthopyroxene. The criss-crossing array of grey bands of serpentine in XP is described as a mesh texture. It is probably a legacy of the fracture pattern in the original olivine, along which water permeated, as was seen in the forsterite marble, M05 (Fig. 2.38). Each 'band' in the mesh pattern goes into extinction,

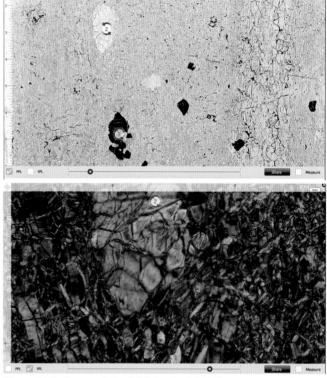

Figure 2.73 Serpentinite (VM GeoLab M18) at low magnification (12mm wide) in PPL and part of the same view at a higher magnification (2mm wide) in XP. For an explanation see the text.

Figure 2.72 Specimen of serpentinite from the Troodos Mountains in Cyprus. The specimen is 11cm wide. *Shutterstock/ www.sandatlas.*

but the bands are not single grains. Each band comprises numerous parallel fibres that go into extinction together.

The hydration reaction for producing serpentine simply requires the addition of water to equal amounts of forsterite and enstatite; these are the two minerals in harzburgite. So to a very good approximation, harzburgite plus water gives serpentinite. The reaction is:

$$2Mg_2SiO_4 \; + \; 2MgSiO_3 \; + \; 4H_2O \; = \; Mg_6Si_4O_{10}(OH)_8$$

forsterite   enstatite   water   serpentine

Asbestos

White **asbestos** is a fibrous variety of serpentine that occurs as **veins** in serpentinite. Veins are fractures in a rock that have gaped open and become filled with new minerals. In an asbestos vein, the fibres of white asbestos are all parallel and appear to have grown out from the walls of the fracture as it dilated (Fig. 2.74). The fibres are flexible and woolly, and they separate from each other easily when they are rubbed.

White asbestos, incidentally, was for many years exploited commercially as a versatile industrial raw material. It was used for insulation, cement slates, brake pads, fireproof blankets, and other products. However, its use is largely outlawed today because exposure to asbestos dust has been implicated in a fatal lung disease known as mesothelioma cancer. Legislation against the use of white asbestos may not, however, be entirely unwarranted. Mesothelioma cancer has been clearly linked to another kind of asbestos, which is a fibrous form of amphibole. The hazard posed by inhaling white asbestos dust is less clear; it is possibly no greater than the hazard of inhaling other forms of silicate dust, which, while harmful and known to cause the lung disease known as silicosis, do not necessarily cause lung cancer, and can be reduced to a tolerable level by taking sensible precautions.

Soapstone

Soapstone is another hydrous variety of metaperidotite, and it occurs in close association with serpentinite. It is a grey-green unfoliated rock made largely from the mineral talc. Talc is a magnesium-rich sheet silicate like serpentine but with the simpler formula, $Mg_3Si_4O_{10}(OH)_2$. Soapstone gets its name because it feels soft and slippery, like soap. Talc is so soft that it rubs off on one's fingers when it is handled; the silicate sheets within it are not well bonded together (see Appendix 2). Soapstone is used as a medium for carving small ornaments and items of jewellery (Fig. 2.75). Talc schist is a foliated variety of soapstone. Soapstone of high purity is pulverized and perfumed and sold as talcum powder.

Figure 2.74 A small piece of serpentinite from the Troodos Mountains in Cyprus. The right-hand side is a 5mm-wide vein of white asbestos seen edge-on.

Figure 2.75 An Inuit carving of a walrus made of soapstone. *Shutterstock/Christopher V Bove.*

2.6.4 Carbonate-bearing metaperidotite
Talc-magnesite rock

GeoLab thin section M27 is a variety of soapstone that contains the carbonate mineral magnesite as well as talc (Fig. 2.76). **Magnesite** is magnesium carbonate, $MgCO_3$. It is seen in the top right quarter of the field of view. Like the other carbonate minerals, calcite and dolomite, it is colourless with high relief in PPL, and has high-order white interference colours (creamy-white with hints of pink and green). It differs from calcite and dolomite in having no deformation twin lamellae. If the rock is examined online, however, twin bands are visible in a carbonate grain at rotation 2, showing that this grain is not magnesite. It is probably dolomite. Magnesite seen online, at rotation 1, changes its relief as the stage is rotated. This behaviour is called **twinkling**, and is common in all carbonate minerals. Talc occupies the central part of the field, as a cluster of randomly orientated flakes showing bright third-order pink and green interference colours in XP. The flakes look a bit like mica flakes.

A simple chemical relationship exists between talc-magnesite rock and serpentinite. One rock can change to the other simply by switching CO_2 for H_2O in the following reaction:

$$Mg_3Si_4O_{10}(OH)_2 \cdot 3Mg(OH)_2 \ + \ 3CO_2 \ = \ Mg_3Si_4O_{10}(OH)_2 \ + \ 3MgCO_3 \ + \ 3H_2O$$

| serpentine | carbon dioxide | talc | magnesite | water |

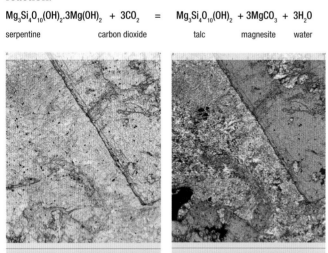

Figure 2.76 Screen shots of talc-magnesite rock (GeoLab, M27) in PPL (left) and XP. Talc (centre) forms a felted mass of flakes with bright colours in XP. Across the top right corner and in the lower left is the carbonate mineral, magnesite, $MgCO_3$. Width of field is 1mm.

If the fluid reacting with peridotite is pure water, serpentinite will be the product, but if the fluid contains CO_2 as well as H_2O, then the product will be a talc-magnesite rock instead.

Anthophyllite-magnesite rock

The last metaperidotite described here (GeoLab, M03) comes from the region of Lewisian gneiss in NW Scotland, close to where the pyroxene granulite (GeoLab, M24) was found. It also contains magnesite, but in place of talc, it has a different hydrous magnesium silicate mineral (Fig. 2.77). This is the pure magnesian amphibole, anthophyllite. **Anthophyllite** is a so-called orthoamphibole (see Appendix 2) and has the formula $Mg_7Si_8O_{22}(OH)_2$. It has parallel extinction (check online at rotation 1), and not the inclined extinction

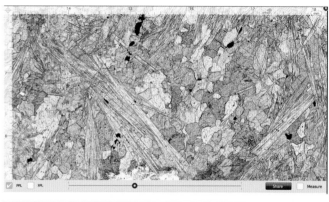

Figure 2.77 Talc-anthophyllite rock (GeoLab M03) in PPL (top) and XP. Anthophyllite forms long spiky and randomly orientated prisms; magnesite forms the rest of the rock. Width of field is 5mm.

of colourless clinoamphiboles like tremolite. It forms long, colourless prismatic crystals. In XP they have bright interference colours, and might be mistaken for muscovite, but they lack the speckled appearance seen in muscovite when it is close to extinction. The random orientation of the prisms indicates that they grew in the absence of directed stress.

The magnesite grains have polygonal outlines. In PPL some show high relief and some low relief. This is because, as in the previous rock, magnesite displays twinkling (this can be seen at rotation 1).

2.7 Summary of metamorphic minerals and protoliths

2.7.1 Minerals and protoliths on an ACF triangle

In this chapter, with the descriptions of metamorphic rocks from the six protoliths, all two-dozen or so of the minerals promised in chapter 1 have been introduced. To summarize their compositions, the ACF diagram is revisited. All the minerals are shown on it (Fig. 2.78). The names of hydrous minerals, i.e. those with (OH) in their formulae, are highlighted in green, and the names of the three carbonates are highlighted in blue. All other minerals are anhydrous. Three minerals, quartz, Na-plagioclase, and K-feldspar have their names listed to one side of the triangle because they have no Ca, Fe, or Mg, and any Al they contain is balanced exactly by (Na + K). The full chemical compositions of the minerals are listed in Appendix 2 (Table A2.1).

Not only does the ACF diagram portray mineral compositions, but it also helps with visualizing the chemical compositions of the protoliths, which are shown on a separate ACF diagram in Figure 2.79. The minerals developed from a particular kind of protolith are grouped within and around the compositional field of that protolith. Thus, pelitic compositions spread from the 'A' corner down the 'A–F' edge, marble compositions spread out from the 'C' corner, metaperidotite compositions all lie close to the 'F' corner, and metabasite

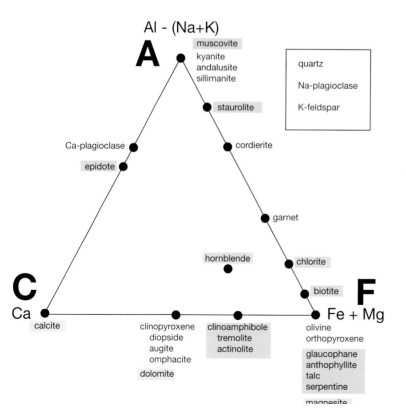

Figure 2.78 ACF triangle showing the positions of the important metamorphic minerals. Names highlighted in green are hydrous minerals, and those in blue are carbonates.

compositions, as already shown, fall near the middle of the triangle.

The diagram does not handle metapsammites or metagranites well, because the minerals in these rocks are mainly quartz, Na-plagioclase and K-feldspar. These three minerals fall outside the triangle, and so, therefore, do the compositions of the metagranites and metapsammites made from them. These last two rock types, nevertheless, can be imagined as being located at separate compositional 'corners' above the plane of the triangle, with metagranites joined to the metabasites via intermediate igneous rocks, and with quartzite and metapsammites joined to the metapelites via semipelitic compositions (Fig. 2.79).

Figures 2.78 and 2.79 together provide a useful *aide memoire* for dealing with unknown metamorphic rocks. Once a rock's minerals have been identified, then, with the help of the ACF triangle (Fig. 2.78), the approximate overall composition of the rock can be estimated, and the protolith can be inferred.

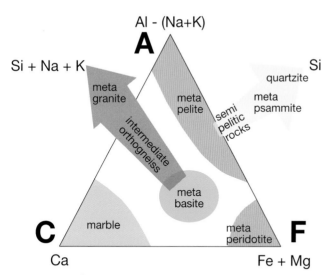

Figure 2.79 ACF triangle showing the compositional fields of four of the six protoliths, with the other two 'floating' above the triangle at the ends of the arrows shown. Metagranite is on an extension from the green metabasite field via intermediate igneous rocks, and quartzite and metapsammite are on an extension from the metapelite field via rocks with semipelitic compositions.

2.7.2 Where do the six protoliths come from?

Peridotite was said in chapter 1 to be like a 'mother rock' from which, ultimately, the other five protoliths are derived. To demonstrate its parental role and the line of descent to its offspring, a flow chart with numbered steps is presented in Figure 2.80. This diagram is the so-called **rock cycle** for the continental crust, modified here by adding partial melting of the mantle. It summarizes all the geological processes described in the book so far.

The standard rock cycle outlines how rocks in the continental crust are changed from one kind to another, or 'recycled'. It looks complicated (Fig. 2.80) but, taken step by step, it is logical and simple to follow. In the continental crust it has a main circuit (thick black line in Fig. 2.80), and several short-circuits. A convenient starting point is the **weathering** and erosion of rocks exposed at the surface (number (1) in Fig. 2.80). Weathering divides rocks into soluble and insoluble fractions, which are transported (2), mostly by flowing water, to lower levels. Soluble elements like K, Na, Ca and Mg are mainly carried in solution, while elements like Fe, Al and Si are mainly moved along as flakes of mud and grains of sand. Either way the material ends up accumulating (3), usually on the seabed, as sand and mud, and also as calcium carbonate shelly material from organisms that extract soluble Ca and CO_2 from seawater. The debris becomes buried and compressed beneath further accumulations of sediment. As it gets deeper it also gets warmer and turns into hard sedimentary rocks through the process of **lithification** (4). Sand becomes sandstone, mud becomes mudstone or shale, and calcium carbonate shells become limestone.

With deeper burial and more intense heating, assisted by orogenesis, metamorphism begins (5) and the sedimentary rocks change into rocks such as schist, marble and quartzite, which are the subject of this book. With yet further heating, schist partially melts (6), becoming migmatite, and granitic **magma** is a by-product. This magma gathers at depth, then rises through the crust because of its low density (7). The main cycle then divides: the rising magma may either travel directly to the surface and erupt as volcanoes (8), or it may form plutonic intrusions (9) that will, later, be brought to the surface by uplift and erosion (10). Either way, the granite and its volcanic equivalent will be subjected to weathering and erosion, and the cycle will begin again.

Four 'short-circuits' to the main loop are shown by thin

black lines, two of them dashed, in Figure 2.80. Firstly, sedimentary rocks can return directly to the surface by uplift and erosion (11) without becoming metamorphosed. Secondly, metamorphic rocks can be exhumed (12) without ever reaching the stage of partial melting. Thirdly, **plutonic igneous rocks** like granite can remain beneath the surface and become metamorphosed directly (13). Fourthly, volcanic rocks can similarly become buried and metamorphosed directly, without being weathered and eroded (14).

The rock cycle in the continental crust accounts for the formation of four of the protoliths, namely sandstone, shale, limestone and granite. To show how these four have descended from peridotite, and to account for the protolith *basalt*, mantle processes have been added. They are shown as green lines. Decompression melting of mantle peridotite produces basaltic magma (15), which rises into the continental crust (16) where it follows a similar path to granitic magma, via plutons or volcanoes, to the surface. At the surface, it is weathered, eroded and transported in

the normal way. Thus, basaltic magma can be thought of as a primary feedstock for the crust, and its parent rock, mantle peridotite, can be regarded as the 'mother rock' for all five other protoliths.

Two further pathways are shown in Figure 2.80. Firstly, some basaltic magma from the mantle ponds at depth in the continental crust (17) where it contributes heat for metamorphism and partial melting, making more granitic magma. Secondly, the small amount of peridotite within the continental crust is evidently added mechanically, as solid chunks (18), e.g. as ophiolite slabs, during plate convergence.

To conclude, metamorphic rocks lie in a box at the confluence of four inward streams. One brings in the three sedimentary protoliths, two streams (from plutons and from volcanoes) deliver the igneous protoliths, and one introduces peridotite from the mantle.

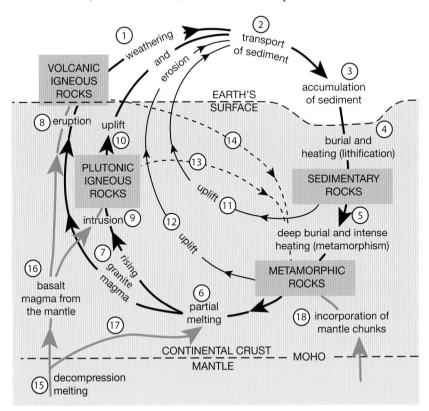

Figure 2.80 The rock cycle in the continental crust (black lines) and its links with the mantle (green lines). For an explanation see the text.

3 Interpreting mineral changes and textures

This chapter addresses questions raised by the observations made in the last chapter. It explores in greater depth the origins of some of the mineral assemblages and textural features of metamorphic rocks. The chapter is in two parts. The first covers the stability of mineral assemblages, the uptake and loss of fluids, and partial melting. The second part deals with metamorphic textures.

3.1 Mineral stability, fluids, and partial melting

3.1.1 What is the meaning of stability?

In chapter 1 it was stated that a change in pressure and temperature beneath the surface causes existing minerals to become unstable together, and that metamorphism occurs when they react chemically with each other to produce new minerals that are stable together under the new pressure and temperature conditions. The drive to attain **stability** between minerals lies at the heart of metamorphism. But what does *stability* mean?

Stability is a state of minimum energy. The idea of minimum energy can be visualized by considering **potential energy**, which is the energy an object has due to its elevation. A boulder perched on a ledge on the steep side of a valley (Fig. 3.1) has excess potential energy and so it is not stable; it can lose its potential energy by being pushed off the ledge so that it rolls to the bottom of the valley. When it can roll no further, its potential energy will have reached a minimum and it will be stable.

Incidentally, if the boulder were not pushed, it would remain forever on the ledge, *appearing* to be stable. The term **metastable**, rather than unstable, is preferred for this situation. A metastable state is where something will remain unstable, and unchanged, unless given an initial push.

Minerals possess a kind of energy that is different from potential energy. It is known as chemical energy. Rather than being linked to elevation, chemical energy is related to the internal arrangement of the atoms within each mineral. It is formally known as **Gibbs energy** and is given the symbol 'G'. A metamorphic rock is stable when it has lost all the Gibbs

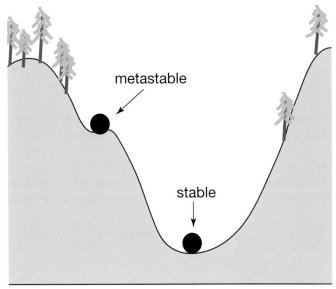

Figure 3.1 Cartoon of metastable and stable states of boulders in a valley.

energy it can lose, i.e. when the combined G of all its minerals is at its lowest possible value. A stable mineral assemblage is said to be in **equilibrium**. Stability and equilibrium are synonyms.

The Gibbs energy in a mineral varies with pressure and temperature. To visualize how it varies, the Gibbs energies for each of the three polymorphs of Al_2SiO_5, kyanite, andalusite and sillimanite, are shown schematically as curved surfaces in a block diagram, Figure 3.2. In this diagram, G increases upwards, with P and T on the two horizontal axes. At any chosen combination of pressure and temperature, the curved surface with the lowest G corresponds to the stable mineral, and the other two minerals are metastable. Metastable sillimanite can exist, for example, in the kyanite field, but having higher G than kyanite, it will lose energy and change to kyanite if, like the perched boulder on the side of the valley, it is given an appropriate 'push'.

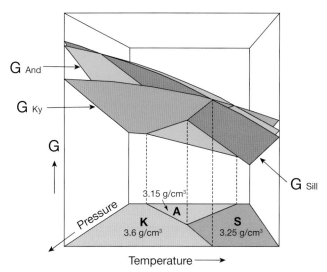

Figure 3.2 Schematic diagram showing Gibbs energy (G, vertical axis) as it varies with pressure and temperature for kyanite (K, blue curved surface), andalusite (A, green surface) and sillimanite (S, red surface). The stability field for each mineral is the P–T range for which it has the lowest G of the three, and is projected onto the base, which is the P–T diagram for Al_2SiO_5. Densities of the polymorphs are also shown.

Where the lowest surfaces intersect, the two minerals at the intersection are equally stable. The intersection line is the boundary between their stability fields. All three stability fields are projected onto the base of the block diagram to give the Al_2SiO_5 P–T diagram shown in chapter 2 (Fig. 2.32). The three surfaces intersect at a unique point where all three polymorphs are in equilibrium together. This is called the triple point, and is located at a pressure close to 4kbar and at a temperature near 500°C.

The variation of G with pressure reflects density. Pressure favours a higher density, so density increases (see the values of density on the base of Fig. 3.2) when, with an increase in pressure, andalusite changes to kyanite or sillimanite, or when sillimanite changes to kyanite.

The principle of stability, and the idea of a surface with minimum Gibbs energy, can be extended from Al_2SiO_5 to a rock with any composition made from any number of minerals. Every stable metamorphic rock at a particular pressure and temperature consists of minerals which,

together, have the lowest possible Gibbs energy for the rock's composition. For example, in the P–T diagram for metabasites (Fig. 2.61) each field shows the P–T area where one kind of metabasite, e.g. amphibolite, has lower G than any of the other kinds of metabasite.

A rock is metastable when it persists as a set of minerals for which G is not as low as it could be. Most metamorphic rocks at the Earth's surface are, in fact, metastable because, having equilibrated at depth, they were unable to continue adjusting to a stable state during their cooling and exhumation.

3.1.2 How was the Al_2SiO_5 diagram obtained?

The Al_2SiO_5 diagram (Fig. 2.32) was obtained by a series of experiments using a bench-top device, called a piston-cylinder apparatus (Fig. 3.3), which simultaneously generates very high pressures and temperatures in a controlled way.

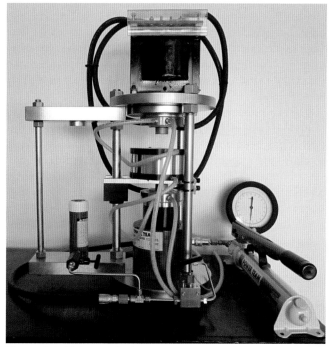

Figure 3.3 Piston cylinder apparatus used in experimental petrology. The piston is forced into the vertical heated cylinder in which the sample is located, and the hydraulic pressure displayed on the circular gauge can be converted to the pressure inside the cylinder. *https://en.wikipedia.org/wiki/ Piston-cylinder_apparatus#/media/File:QUICKpress.jpg*

54

To locate the field boundaries of kyanite, andalusite and sillimanite, a powdered mixture of *all three* polymorphs is sealed in a gold capsule, and embedded in a ductile solid medium, such as powdered talc, inside a strong metal cylinder. The cylinder is closed at one end and a piston is pushed by a hydraulic jack into the other end, compressing the sample to a known pressure. The assembly is heated by an internal electrical resistance heater and the temperature is monitored by a thermocouple inserted into the cylinder through a tiny hole. The sample is kept at a known high pressure and temperature for long enough for it to equilibrate, or at least to move decisively in the direction of equilibrium (i.e. towards the bottom of its Gibbs energy valley). This may take weeks or months. The sample is then quenched (cooled rapidly), and the content of the gold capsule is examined, usually by SEM or XRD (see Appendix 4). The polymorph with the lowest G at the chosen P–T conditions can be identified, because it will have increased in amount at the expense of the other two. Such experiments are carried out multiple times, covering different combinations of pressure and temperature, so that the positions of the three boundaries can be tightly constrained.

The experimental investigation of the stability of rocks and minerals is a branch of research called **experimental petrology**. It has been applied to various 'rock' compositions to establish the equilibrium boundaries for many minerals and sets of minerals. These boundaries correspond to chemical reactions between the minerals on either side. The results of experimental petrology make it possible to be fairly sure about the P–T values stated earlier for metamorphic grades, zones and facies (e.g. Fig. 1.6 and Fig. 2.61). The results for the stability of a few selected minerals are discussed in chapter 5.

3.1.3 What kinds of metamorphic reaction produce water?
Metamorphic reactions involving loss of water are common. Strictly speaking, they involve loss of *aqueous fluid*, which can be either water or steam, depending on the pressure. However, for simplicity the word *water* is used here for all aqueous fluid, whether it is steam or water.

Loss of water results from two kinds of reaction. One kind is represented by the reaction between muscovite and quartz, and the other is represented by the breakdown of biotite in medium-grade schist.

Taking the first kind of reaction, muscovite and quartz are common together in metapelites in all six Barrow zones, but in the high-temperature half of the sillimanite zone, where migmatite develops, muscovite disappears by reacting with quartz; K-feldspar and sillimanite appear in their place (see Fig. 2.28). Muscovite has a fixed composition (i.e. it has no atomic substitution), and it loses its water all in one go when it reacts with quartz.

The breakdown reaction can be written:

$$KAl_2(Si_3Al)O_{10}(OH)_2 \; + \; SiO_2 \; = \; KAlSi_3O_8 \; + \; Al_2SiO_5 \; + \; H_2O$$

muscovite　　quartz　　　K-feldspar　sillimanite　water

This reaction has been investigated experimentally and the stability field for muscovite plus quartz is shown by a line in Figure 3.4. This line can be regarded, like the field boundary between kyanite and sillimanite, as a line along which muscovite, quartz, K-feldspar, Al_2SiO_5 and H_2O coexist in equilibrium – a line where muscovite plus quartz together have exactly the same value of Gibbs energy as K-feldspar, sillimanite and H_2O together. At lower pressures, andalusite is present instead of sillimanite.

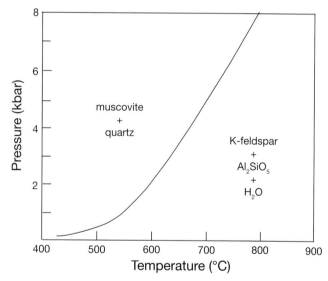

Figure 3.4 Pressure–temperature diagram showing the experimentally determined reaction between muscovite and quartz. The line shows where muscovite, quartz, K-feldspar, Al_2SiO_5 and water are all stable together. Al_2SiO_5 is either sillimanite or andalusite, depending on the pressure.

Incidentally, the equilibrium line in Figure 3.4 is noticeably curved at pressures below about 2kbar. The reason is that H_2O is a compressible gas which, at these pressures, expands enormously as pressure falls. Thus, at very low pressures the combination of Al_2SiO_5, K-feldspar and H_2O (the reaction products) occupies a much greater volume than the combination of muscovite and quartz. The huge difference in volume makes the reaction very sensitive to a change in pressure, and so the slope of the line at low pressures is quite gentle. Almost all dehydration reaction lines are curved like this at low pressures.

Turning to the second kind of reaction that releases water, biotite, in contrast to muscovite, does not have a fixed composition but has substitution of Fe for Mg. As a result, biotite reacts gradually with other minerals, over a range of temperatures, as it is heated and loses H_2O. With each incremental rise in temperature a tiny amount of H_2O is released, and the biotite is reduced very slightly in amount, and has a slightly lower Fe/Mg ratio than before. The reaction can be written:

biotite₁ (higher Fe/Mg) + reactants = biotite₂ (lower Fe/Mg) + products + H₂0

The reaction is shown in a schematic way in Figure 3.5. Water is released continuously at a slow rate as the temperature rises. The reaction is described as a **continuous**

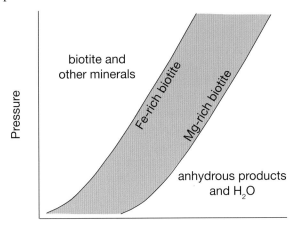

Figure 3.5 P–T diagram showing, conceptually, the dehydration of biotite by reaction with other minerals during heating of a pelitic rock. Biotite breaks down continuously through the shaded interval. As the amount of biotite decreases to zero, its composition becomes steadily more Mg-rich.

dehydration reaction, while the reaction between muscovite and quartz (which happens at a fixed temperature for a given pressure) is described as **discontinuous**. Other hydrous minerals with Fe and Mg, such as chlorite and amphiboles, lose their water by continuous dehydration, like biotite, and become more magnesium-rich as they decline in volume.

3.1.4 How much water is tied up in metamorphic minerals?

About half of the two-dozen or so metamorphic minerals listed in Table A2.1 (Appendix 2), and shown in Figure 2.78, are hydrous. These minerals are ranked by water content in Table 3.1. Serpentine and Mg-chlorite are the most hydrous, with about 13% by weight of H_2O in each. The other sheet silicates (talc, muscovite and biotite) each contain roughly 4% by weight of H_2O. Hornblende and the other amphiboles carry about 2% H_2O, as does epidote. Staurolite has about 1% H_2O.

The percentages in Table 3.1 were calculated from the formula of the mineral and the atomic weights of the elements. For example, the formula of muscovite, $KAl_2(Si_3Al)O_{10}(OH)_2$, can be re-written as $KAl_3Si_3O_{12}H_2$. The atomic weights are 39 for K, 27 for Al, 28 for Si, 16 for O, and 1 for H. The total weight of one complete formula is therefore 398 units of atomic weight, being the sum of 39 for K, 3 by 27 for Al, 3 by 28 for Si, 12 by 16 for O, and 2 by 1 for H. With two H atoms, there is one lot of H_2O present in the formula, and it weighs 18 units. So the weight percentage of H_2O is $(18/398) \times 100$, which is 4.52%.

The weight percent of water in minerals with Fe–Mg substitution varies with the ratio Fe/Mg. Pure Mg biotite, for example, has a formula weight of 417, of which water, weighing 18 units, is 4.3%. Pure Fe biotite has a higher formula weight, 511, so the water content is lower, at 3.5%.

3.1.5 How does the water content in metapelites change with grade?

Prograde metamorphism leads to the loss of water. The decreasing percentage of water tied up in a representative metapelite, going through the Barrow zones, is illustrated in Figure 3.6. The values are based on the figures in Table 3.1, and the amount of each kind of mineral in the rock. At low grade, for example, a muscovite-chlorite-schist with 30%

56

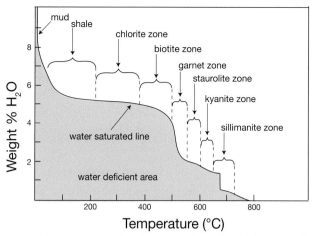

Figure 3.6 Diagram showing the approximate weight % of H_2O in a representative metapelite as metamorphic temperature increases through the Barrow zones. Loss of water follows a curved line, except for a hiatus in the middle of the sillimanite zone where it drops abruptly by about 1%. This drop marks the reaction between muscovite and quartz.

Table 3.1 Water contents, in weight %, of hydrous minerals.

Name	Formula	Wt.% H_2O
Serpentine	$Mg_6Si_4O_{10}(OH)_8$	13%
Mg chlorite	$Mg_5Al_2Si_3O_{10}(OH)_8$	13%
Fe chlorite	$Fe_5Al_2Si_3O_{10}(OH)_8$	10%
Talc	$Mg_3Si_4O_{10}(OH)_2$	4.8%
Muscovite	$KAl_3Si_3O_{10}(OH)_2$	4.5%
Mg biotite	$KMg_3Si_3AlO_{10}(OH)_2$	4.3%
Fe biotite	$KFe_3Si_3AlO_{10}(OH)_2$	3.5%
Glaucophane	$Na_2Mg_3Al_2Si_8O_{22}(OH)_2$	2.4%
Mg hornblende	$Ca_2Mg_4Al_2Si_7O_{22}(OH)_2$	2.2%
Fe hornblende	$Ca_2Fe_4Al_2Si_7O_{22}(OH)_2$	1.9%
Epidote	$Ca_2Al_3Si_3O_{12}(OH)$	2.0%
Staurolite	$Fe_2Al_9Si_4O_{23}(OH)$	1.1%

chlorite, 35% muscovite and 35% quartz contains just over 5% H_2O. This is calculated roughly as 30% of 12% for chlorite, plus 35% of 4.5% for muscovite, plus 35% of zero for quartz. With increasing grade, the water is lost gradually as chlorite slowly disappears. Only about 2% of water remains in the garnet zone, where a pelitic schist typically contains about 25% each of muscovite and biotite. With further increase in grade, more water is lost as biotite disappears gradually, but water is released in a gush when the reaction line for muscovite plus quartz (Fig. 3.4) is crossed. If, for example, the muscovite is 20% of the rock then, when it goes, the water in the rock will drop by almost 1% (Fig. 3.6).

The upper boundary to the shaded area in Fig. 3.6 is called the water-saturated line. It is the maximum amount of water that can be bound up as hydrated minerals in the representative metapelite at any given temperature. Conditions above the line are never met because excess water would move away to lower pressures. The grey area, below the water-saturated line, can be described as **water-deficient**. No free water can exist here. If free water were present, it would immediately become consumed by producing hydrous minerals.

3.1.6 How are stable mineral assemblages in metapelites preserved?

A general question in metamorphism is how equilibrated mineral assemblages, preserved in medium- and high-grade rocks, do not continue to equilibrate during cooling as they return to the surface, but become metastable. Figure 3.7 suggests a possible answer to this question in the context of metapelites. During heating through the successive Barrow zones, the discontinuous breakdown of chlorite and biotite means a perpetual small amount of H_2O will exist along grain boundaries. It is believed that this water promotes on-going equilibration among the minerals, which is shown in Figure 3.7 by a red line and arrow. It is called **prograde** metamorphism because it is associated with increasing grade. However, once the peak temperature has passed and cooling sets in, equilibration will probably stop abruptly. This is because any tiny amount of remaining H_2O will quickly be mopped up by back reactions, leaving absolutely no free water along grain boundaries to promote further equilibration. The cooling path is shown in blue in Figure 3.7. The water-deficient rock will generally stay unreactive and metastable during its journey back to the surface.

The water released during heating, incidentally, is presumed to make its way towards the surface, to regions of lower pressure. It does not always get to the surface, but can become trapped beneath an impermeable barrier layer. An example of this was seen in the Russian super-deep borehole on the Kola Peninsula mentioned in chapter 1, where trapped water (presumed to have been released by

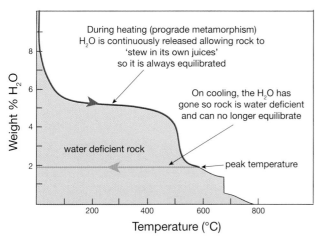

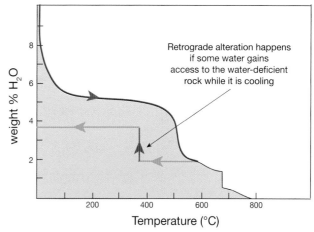

Figure 3.7 The same diagram as shown in Figure 3.6 but with a red arrow and line (the water-saturated line) showing the path followed by metapelites during prograde regional metamorphism. With water being steadily released, the rock continues to equilibrate up to the peak temperature. Thereafter, the rock becomes water-deficient (blue line) and is unable to equilibrate further, so the peak mineral assemblage is preserved.

Figure 3.8 The same diagram as Figure 3.7 but with retrograde alteration added. Addition of some water during cooling (vertical red arrow) leads to aqueous alteration (called retrograde change), with new hydrous minerals growing along cracks and grain boundaries until all available water is consumed. If excessive water were to be added, the vertical red arrow would reach the top of the water-deficient area and the rock would become completely re-equilibrated.

prograde metamorphism) was discovered in fractured rock at a depth of 7km.

3.1.7 Retrograde alteration and complete re-equilibration

Water-deficient, metastable rocks do not always survive in a pristine state back to the surface. Some metamorphic changes can, and often do, affect them; these are the so-called **retrograde** metamorphic changes, i.e. changes associated with decreasing grade, mentioned in chapter 2.

One common kind of retrograde change is caused by the late ingress of a limited amount of water. This added water is represented by the vertical red arrow in Figure 3.8. As soon as the limited supply of water is consumed (by the formation of new hydrous minerals) the rock will again be water-deficient and unreactive. Examples of retrograde hydration include the serpentine-filled cracks within forsterite (Fig. 2.38), and the borders of hornblende and plagioclase between garnet and omphacite in eclogite (Fig. 2.55).

What is the source of this late-injected water? One possibility is that it comes from prograde reactions deeper in the orogenic root, particularly where medium- or high-grade

rocks get shunted upwards by tectonic forces, and emplaced over low-grade metamorphic or sedimentary rocks that are getting hotter.

If the amount of late-injected water were sufficiently high, the vertical red arrow in Figure 3.8 would extend up and meet the water-saturated curve. The rock would then contain all the water it could manage and, with water again permeating along grain boundaries, it would develop a new stable mineral assemblage, completely replacing the original one.

3.1.8 Water in metabasites and metaperidotites

The water-deficient area for a representative metabasite is shown in Figure 3.9. It is like that for a metapelite, with much of the water being lost at around 500°C as chlorite breaks down. However, the metamorphic history of a metabasite differs fundamentally from that of a metapelite. Whereas metapelites start off as wet mud, and steadily *lose* water as they are metamorphosed, metabasites start out bone dry (basalt is made of the anhydrous minerals augite and plagioclase). Thus, metabasites with hydrous minerals will actually have *gained* water during metamorphism. The basic igneous

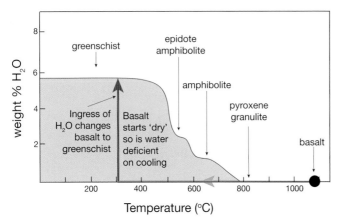

Figure 3.9 The water-deficient area (grey) for metabasites at different temperatures. The protolith, e.g. basalt, is anhydrous, so metamorphism first requires the introduction of water. The vertical red arrow shows that almost 6% by weight of water is needed to convert dry basalt to greenschist. Once hydrated, and on subsequent heating, the rock will undergo normal prograde metamorphism and lose water, just as metapelite does.

protolith, on cooling below about 800°C, automatically finds itself as a water-deficient rock, and will remain unchanged if water fails to gain access, even if the surrounding rocks become metamorphosed. An example of dolerite that has survived medium-grade metamorphism is shown in Figure 3.10. It comes from a 2400**Myr** (million year) old dyke intruding so-called Lewisian gneiss in NW Scotland. The dyke and country rock were heated to amphibolite-facies conditions 1700Myr ago, but the dolerite shows no sign of changing to amphibolite because it remained water-deficient.

Usually, however, water makes its way into the rock and hydrates it, for example to blueschist, greenschist or amphibolite, depending on the ambient P–T conditions. If the temperature then rises, the rock will undergo normal prograde metamorphism, and the water released by dehydration reactions will promote continuous equilibration, just as it does in metapelites.

Turning to peridotite, this protolith is also anhydrous. Figure 3.11 shows that it can absorb a huge amount of water – about 13% by weight – when it is converted to serpentinite. In general, the temperature must be below about 500°C for serpentinite to form. If serpentinite later becomes heated above about 500°C, it will lose its water through a series of reactions,

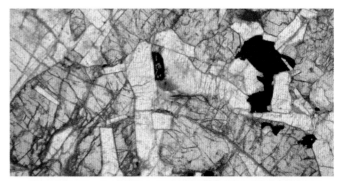

Figure 3.10 Thin section of dolerite in PPL and XP. In PPL the colourless grains are Ca-plagioclase, and the pale brown ones are augite. This rock, from a dyke in NW Scotland, has experienced amphibolite-facies conditions but, being water-deficient, it survived intact. Width of field is 3mm.

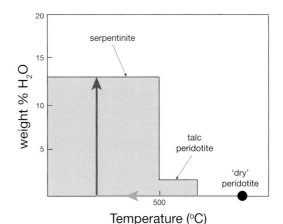

Figure 3.11 The water-deficient area (grey) for metaperidotite, highly simplified. Serpentinite contains a huge 13% by weight of H_2O.

simplified in Figure 3.11 to just two steps, namely the formation of olivine-talc rock and, at a higher temperature, anhydrous peridotite.

The water that hydrates basic igneous rocks and peridotite may come from neighbouring pelitic rocks that are undergoing prograde metamorphism, or, as suggested above in the context of retrograde changes generally, it may be derived from cool, water-bearing rocks that become deeply buried beneath hot, dry rocks during orogenic convergence. A different, and probably more important, source of water is the groundwater driven by thermal **convection** which causes hydrothermal metamorphism, as described in chapter 1 (Figs 1.10 and 1.11). This process is thought to transfer a truly enormous amount of water into the newly-formed oceanic crust. The water will end up in greenschist and amphibolite and, if the circulation is deep enough, in serpentinite (derived from the underlying harzburgite). In addition, water will be taken up at lower temperatures in zeolites and hydrated volcanic glass. When the oceanic crust eventually becomes subducted, this water will be carried deep into the mantle, to be released in quantity as the temperature exceeds about 500°C. The consequences of this deep release of water are discussed in chapter 4.

3.1.9 Fluids other than H_2O

Only one fluid, pure H_2O, has been considered so far in this chapter. Two other fluids are now acknowledged, albeit briefly – salty water, and CO_2.

Metamorphic water is probably more often salty than pure, as is exemplified by the beautiful salt crystal in the fluid inclusion figured in chapter 1 (Fig. 1.3). Salty water has an interesting implication. If retrograde hydration involves salty water, and if the retrograde minerals have no capacity for taking in the chemical elements in the dissolved salts (mainly Na and Cl), then a small amount of extremely concentrated salty water may be left behind. This could, potentially, promote equilibrium in rocks that are otherwise water-deficient.

Carbon dioxide is also important. It is released during the prograde metamorphism of impure limestone, when silicates like diopside and forsterite are produced. As it migrates upwards it is likely to become mixed with H_2O from neighbouring dehydration reactions.

Carbon dioxide is completely miscible with H_2O under most metamorphic conditions; any fluid composition between pure H_2O and pure CO_2 is possible. The presence of mixed H_2O plus CO_2 fluids can dramatically alter the P–T conditions for metamorphic reactions where H_2O or CO_2 is a product. As was noted in chapter 2, the presence of some CO_2 with H_2O in the fluid can affect the low-grade metamorphism of peridotite, where the addition of pure, or nearly pure, water converts peridotite to serpentinite, whereas addition of water with a small amount (perhaps 10%) of CO_2 present can change the peridotite, instead, to a talc-magnesite rock, or at higher grade, an anthophyllite-magnesite rock.

3.1.10 Partial melting and the origin of migmatite

This section looks at the role of H_2O in the formation of migmatite. As was described in chapter 2, if metapelite becomes hot enough during metamorphism, it will begin to melt. It will not melt completely, but it will *partially* melt over a temperature interval, becoming a kind of slush. This process begins within the sillimanite zone (probably at about 700°C), and results in migmatite, a kind of gneiss with small layers and lenses of once-molten granite in darker surroundings (e.g. Fig. 2.29). Evidently the 'slush' somehow segregates into small volumes of granitic liquid and residual solids which, on cooling, become the leucosome and melanosome, respectively.

Experimental petrology relating to granite has now shown that molten granite requires a temperature of about 950°C, far higher than the 700°C inferred for migmatite. Secondly, the individual minerals in granite (i.e. quartz and feldspar), taken alone, do not melt until even higher temperatures are reached – between 1050°C and 1200°C for feldspar (depending on its composition), and 1700°C for quartz. These observations raise two separate questions. Firstly, how can patches of liquid granite appear in migmatite when the temperature is only about 700°C, some 250°C below the temperature at which granite is observed to melt in experiments? Secondly, how can this experimentally measured temperature for the melting of granite (about 950°C) be so much lower than the melting temperatures of its individual mineral constituents?

To answer the second question first, this behaviour is normal for any mixture of minerals. To explain it, the concept of a **eutectic** (pronounced you-**tek**-tik) needs to be understood. With sufficient heating, a mixture of two or more minerals will commonly *partially melt* (start to melt) at a temperature that is much lower than the melting point of any of the minerals on its own. Moreover, the temperature at which partial melting begins is *fixed* for any particular set of minerals. Also, the chemical composition of the first-formed liquid is fixed, regardless of the relative amounts of each of the minerals. The fixed temperature and the fixed chemical composition of the first-formed liquid are known, respectively, as the eutectic temperature and the eutectic composition.

Eutectic melting can be illustrated using the example of ice crystals and salt (NaCl), familiar to anyone who has seen icy roads being salted in winter (Fig. 3.12)

The NaCl–H_2O 'melting' diagram has four fields (Fig. 3.13) – solid ice and salt, ice with saltwater, salt with saltwater, and saltwater alone. The four fields meet at the eutectic point. On the left side, ice melts at 0°C. On the right side, NaCl melts at

Figure 3.12 Salting roads to clear ice. *Photo courtesy of Minnesota Pollution Control Agency.*

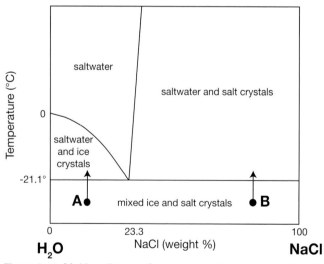

Figure 3.13 Melting diagram for a mixture of ice and salt (NaCl). For an explanation of paths 'A' and 'B' see the text.

about 800°C (well off the scale). Between them, ice and salt together *always* start to melt at -21.1°C (the eutectic temperature) to produce saltwater with 23.3% by weight of NaCl (the eutectic composition). As a simple experiment, one could place two containers into a deep freeze at -25°C, one with abundant ice and little salt (position 'A' in Fig. 3.13), and another with a little ice and abundant salt (position 'B'), and let them warm up to -20°C, just above the eutectic temperature. In both cases saltwater will appear, and in both cases it will have 23.3% NaCl in solution, the eutectic composition. In the first case the eutectic saltwater is mixed to form a slush with a good deal of unmelted ice (in the saltwater and ice field); in the second, a slush is formed from the eutectic saltwater and a lot of undissolved salt (in the saltwater and salt crystals field). If one did the same experiment with a mixture of exactly 23.3% salt and 76.7% ice (by weight), the mixture (which is the same as the eutectic composition) would completely melt at -21.1°C. Importantly, the first liquid to form always has the eutectic composition, regardless of the amounts of ice and salt in the starting mixture.

Turning to rocks, a similar diagram to the ice–salt melting diagram, with four fields, exists for quartz and feldspar, and is shown in Figure 3.14. Quartz alone melts at 1700°C, as mentioned above, and feldspar (taken here as Na-K-feldspar with roughly equal Na and K) melts at about 1050°C. Quartz and Na-K-feldspar together, however, begin to melt at a eutectic temperature of 950°C and the first liquid to form is roughly equivalent to two-thirds feldspar and one-third quartz. This is the eutectic composition. It is also the composition of granite. Thus granite is a rock with a eutectic composition. Granitic magma is the first liquid to appear when any rock containing some quartz and some feldspar, in any ratio, begins to melt. Most metamorphosed semipelitic and pelitic rocks contain quartz and feldspar, and much of the granite in the upper continental crust may come from such sources.

The other question asks how the granite lenses in migmatite are able to be molten at only 700°C, when the granite eutectic temperature is 950°C. The answer is linked to the fact that partial melting in migmatites is linked to the presence of water. Once the pressure is high, the eutectic temperature for quartz, feldspar *and water* is very different from, and much cooler than, the eutectic temperature for quartz and feldspar alone, without water. Figure 3.15 shows how the two different eutectic temperatures for granite – one for granite in the

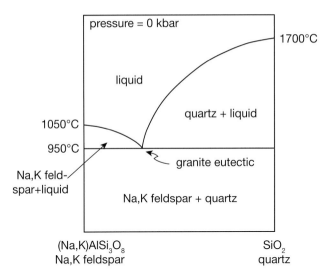

Figure 3.14 Simplified melting diagram for a rock made of Na,K feldspar and quartz. The granite eutectic point is labelled. The first liquid to appear as the temperature rises will be granitic, regardless of the relative amounts of quartz and Na,K feldspar in the rock being heated.

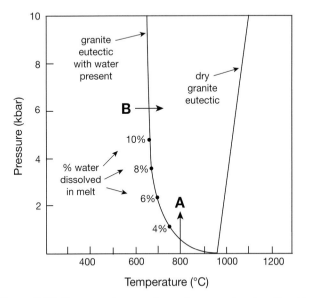

Figure 3.15 Pressure–temperature diagram showing the 'dry' granite eutectic, and the eutectic for granite where H_2O is present. For an explanation see the text.

presence of water and the other for dry granite – change with increasing pressure. The lines on this diagram show that both eutectics start at 950°C when the pressure is very low, but with increasing pressure the dry eutectic slopes forwards, while the 'wet' eutectic slopes backwards, then curves upwards to become almost vertical at about 650°C. Why is this so?

The dry eutectic line in Figure 3.15 slopes forwards for the same reason that the line for the melting of mantle peridotite slopes forwards (see section A1.2.1 in Appendix 1). It is because, in both cases, the liquid is less dense than the solids, and it lies behind the important process of partial melting by decompression.

The wet eutectic line slopes backwards because at a high pressure water can dissolve in granitic liquid. The approximate percentage by weight of water that can go into granitic magma is shown at four different pressures on Figure 3.15. It reaches 10% at 5kbar, which is a pressure that might be expected for metapelites in the sillimanite zone. The eutectic melting at this pressure results from a reaction between *three* components – quartz, feldspar *and* H_2O – to give granitic liquid with a large amount of dissolved water (wet granite). Since H_2O is a compressible gas, and occupies a large volume at low pressure, the overall density of quartz, feldspar *and* H_2O *gas* at low pressure is much less than the density of wet granite liquid in which the H_2O gas is dissolved. Therefore, an increase in pressure favours 'dense' molten 'wet' granite over a 'less dense' mixture of solid quartz and feldspar grains and water. Put another way, a combination of quartz, feldspar and H_2O at 800°C and zero pressure would start to melt if it were compressed to 1kbar, following path 'A' in Figure 3.15.

A problem with appealing to the wet granite eutectic curve to explain the formation of migmatite is that melting can only happen if sufficient water is present to stabilize the melt. Most metapelites are destined to *start* to melt once the temperature of the 'wet' granite eutectic is exceeded (about 650°C; arrow 'B' in Fig. 3.15) because they contain at least some quartz and feldspar, and they also contain biotite so, as metamorphic grade increases, they are constantly bathed in a small amount of their own water, which is released by continuous dehydration of the biotite. But very little granitic liquid can form at this stage because there is so little water available to stabilize it. The supply of water jumps dramatically, however, at roughly 700°C, when muscovite reacts with quartz, as described above. Metapelite with 25% muscovite will yield about 1% of the rock's weight as H_2O. This will immediately promote the formation of more hydrous granitic liquid. Since the liquid needs 10% of H_2O to be stabilized, and 1% of the rock's weight is suddenly available as H_2O, then, provided enough quartz and feldspar are present, about 10% of the rock will melt and appear as granitic lenses in the migmatite. In summary, it is not the 'wet' granite eutectic that controls migmatite formation; rather it is the release of water through dehydration reactions, such as the reaction between muscovite and quartz, after the 'wet' granite eutectic line has been crossed.

A point of interest regarding migmatite concerns water deficiency. As the granitic liquid in a migmatite cools and solidifies, the water it contains will be released back into the surrounding water-deficient rock, where it has the potential to promote a considerable amount of retrograde hydration.

A final observation is that when granitic liquid migrates away from its source in migmatites, to gather into larger pools and eventually to move upwards through the crust and feed into plutons, the residual, unmelted fraction of the rock, left behind at depth, will be a special kind of metamorphic rock. Having lost granite, it will no longer have the chemistry of its protolith. Residual rocks from metapelites will generally have higher than normal amounts of garnet and Al_2SiO_5. Rocks that have changed in composition as a result of becoming partially melted, and losing liquid, are described as 'melt-depleted', or as melt residues. They probably account for a significant fraction of the metamorphic rocks in the lower part of the continental crust in places where granite plutons are common at the surface today. This idea is, in fact, supported by the presence in basalt in continental regions (such as Carboniferous basalt in central Ireland) of xenoliths of melt-depleted, melanosome-like metapelites presumed to come from the lower crust.

3.2 Understanding metamorphic textures

The second section of this chapter deals with the textures of metamorphic rocks. Texture was defined in chapter 1 as the sizes, shapes and orientations of the grains in a rock, and also their spatial distribution. The section begins (3.2.1) with a concise review of the textures described in chapter 2, noting their more obvious significance. It then goes on to explore in more detail some of the processes that gave rise to these textures.

3.2.1 A review of textures as a record of grain growth, strain, and multistage history

Regarding the sizes of grains in metamorphic rocks, they generally get larger with increasing temperature going, for example, from slate, through phyllite and schist, to gneiss, or, for example, from greenschist, through amphibolite to pyroxene granulite. This size increase is attributed to increasing degrees of grain growth. Also, grains of some minerals, called porphyroblasts, are seen to have grown much larger than grains of other minerals in the same rock. In marked contrast, grains in rocks such as mylonite are mostly tiny (apart from some large, deformed grains called porphyroclasts) as a result of intense strain and mechanical damage.

Regarding their shapes, grains may be **euhedral** (having good crystal faces like, for example, dodecahedral garnet), or **anhedral** (without crystal faces). Anhedral grains are commonly **polyhedral** (multifaceted), so that their shapes in a thin section are polygons. A rock in which all the grains are polygonal in a thin section, and about the same size, is described as granoblastic. Such a texture is seen in the pyroxene granulite shown in Figure 2.51.

Regarding their orientation, the parallel alignment of platy and elongated grains, seen as slaty cleavage and schistosity, is attributed to the growth of grains while subjected to directed stress. The corollary is that randomly oriented grains, as in the andalusite hornfels, and the anthophyllite-magnesite rock (Figs 2.31 and 2.77), grew in the absence of directed stress. Crenulated schistosity (Fig. 2.18) is attributed to the crumpling of schistosity in response to a change in the direction of stress during metamorphism.

Together, the sizes, shapes and orientations of grains can be rationalized as being a consequence of two separate, and competing, processes: grain growth and strain. Grain growth is promoted by high temperatures, and strain is caused by directed stress, as summarized in Figure 3.16. Grain growth without strain, and strain without grain growth, are at the two ends of a spectrum of effects. In the middle, grain growth and strain, in tandem, lead to foliated textures.

With regard to the distribution of grains in a metamorphic rock, an uneven distribution is usually evidence for more than one event in the rock's history. In the amphibolite hornfels (Fig. 2.62), for example, the distribution of hornblende and plagioclase is inherited from the texture of the dolerite protolith, giving a so-called a relict igneous texture. As a different

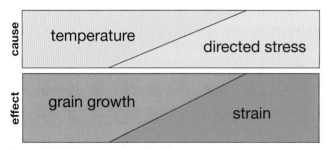

Figure 3.16 Diagram showing that temperature promotes grain growth, while directed stress promotes strain. Most rocks fall into one of three textural categories: those that have textures resulting from grain growth alone (e.g. hornfels), or strain alone (e.g. mylonite), or grain growth and strain together (e.g. schist).

example, in the discussion above on retrograde hydration, both the presence of serpentine along cracks in forsterite and the fine-grained hornblende and plagioclase along grain boundaries in eclogite point to the late ingress of water. In a third example, inclusion trails in garnet porphyroblasts (Fig. 2.26) provide a record of the early stages in the rock's history of deformation and grain growth.

The remainder of this section on textures addresses questions relating only to grain growth and strain. Further examples of how grain distribution provides evidence for a multistage history of metamorphism will be presented in chapters 4 and 5.

3.2.2 What makes grains grow?

Grain growth, which is also called **recrystallization**, is a fundamental metamorphic process. Grains can only grow in size at the expense of other grains that are getting smaller, eventually to disappear altogether. For grains to grow, atomic bonds in the smaller grains must be broken, so that the atoms are released and can then migrate and add on to larger grains. This process is favoured by high temperatures because temperature is a measure of the vigour with which atoms are vibrating and of the frequency with which bonds are being broken. The importance of high temperature is clear from the fact that grain size increases with metamorphic grade.

But while high temperature facilitates recrystallization, it does not explain *what* makes grains grow. To understand that, one needs to return to the idea of stability. It turns out that metamorphism is driven by a strong drive for a rock to

minimize its Gibbs energy, not just in its mineral assemblage, but also in its texture.

The Gibbs energy being minimized with grain growth is energy associated with the surfaces of grains, called surface energy. Surfaces are untidy places as far as atoms are concerned. Each atom on the boundary between two adjacent grains can only be bonded neatly to one of those grains. It will be out of alignment with the regular pattern of atoms in the adjoining grain (Fig. 3.17). Moreover, misfit atoms, too large or too small to be accepted by either grain, often end up stranded at the interface between two grains. This untidiness of grain boundaries gives rise to excess surface energy. The excess will be reduced if the total surface area is reduced, and this obviously happens with grain growth, when many small grains are replaced by a few large ones (Fig. 3.18).

Today it is possible to watch grain growth in real time, not in rocks but in metals, where the principles are the same but the atoms can move around much more easily. A special heated specimen stage inside a scanning electron microscope is used, and a succession of time-lapsed images is taken. Such grain growth can be viewed online at: https://www.youtube.com/watch?v=Cy_rYNc0UAY

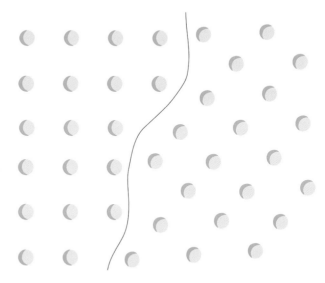

Figure 3.17 Cartoon depicting atoms at a boundary between two grains with different orientations. Atoms on the boundary cannot be bonded neatly to both grains, which results in a small excess of Gibbs energy, called surface energy.

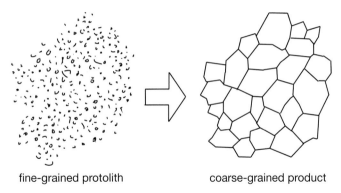

fine-grained protolith   coarse-grained product

Figure 3.18 Cartoon depicting grain growth, before and after, for a rock made from just one kind of mineral.

In this youtube video a fine-grained piece of copper metal is heated, and new large grains can be seen growing at the expense of the small grains.

Two other videos provide insights into the way grains grow. In one of them Sir Lawrence Bragg, the Nobel Laureate in Physics who worked out how to interpret X-ray diffraction by crystals (see Appendix 4), is delivering the 1952 Royal Institution lecture, using rafts of equal-sized soap bubbles as a proxy for crystals, in which the bubbles are atoms: https://www.youtube.com/watch?v=UEB39-jlmdw.

The second video is a gem, a model in clarity of presentation, again relating to grains in metals but highly relevant to metamorphic textures: https://www.youtube.com/watch?v=uG35D_euM-0.

3.2.3 Is time, like temperature, a factor in grain growth?

While the correlation between metamorphic grade and grain size shows that temperature influences grain size, it is not the only influence. An additional factor is the *duration* of high-temperature conditions, or time. The importance of time is self-evident. It takes a long time for grains to grow to a large size. As noted above, strong chemical bonds first have to be broken so that atoms in the small, shrinking grains can move to the growing grains. But atoms are not easily able to tell which grains are good bets to join and which are not, so the process is one of trial and error, with many failed attempts by grains to grow before a few grains eventually prevail at the expense of the others.

The importance of time can perhaps be demonstrated by comparing pyroxene granulite in a regional metamorphic setting with pyroxene granulite (as hornfels) in a contact setting. Both have the same mineral assemblage, so presumably formed at roughly the same temperature. In the first case the grains are large, whereas in the second case they are small (Fig. 3.19). A plausible reason for the difference is that grains continue to grow for a very long time in a regional setting, whereas in a contact aureole the brief duration of high temperatures gives them little time to grow.

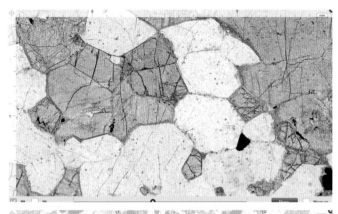

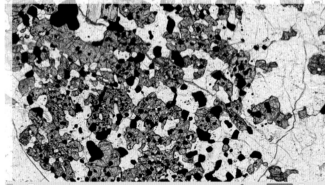

Figure 3.19 Thin sections in PPL of two contrasting pyroxene-plagioclase rocks. Upper image is a pyroxene granulite (M24) from a regional metamorphic area where grain growth lasted for millions of years. Lower image shows a pyroxene hornfels from a contact aureole (VM UK collection) where grain growth was short-lived. Both images are 2.5mm wide.

3.2.4 Does fluid have a role in grain growth?

Even at a high temperature and with ample time, it seems that grains will not grow without the help of a third factor – the presence of fluid along the grain boundaries. The importance of fluid was alluded to in section 3.1.6 above, in the context of the prograde metamorphism of metapelites. Fluid facilitates grain growth because it assists the breaking of the strong chemical bonds in the smaller grains, allowing them to dissolve. Once in solution, the atoms can move freely to a place where they can come out of solution and add to the surfaces of growing grains. The fluid in this case, by permitting easy movement of atoms without being involved in chemical reactions, is said to act as a catalyst.

An excellent field example that highlights the importance of fluid is shown in Figure 3.20, where an outcrop of pyroxene granulite gneiss that was formed about 1000Myr ago is traversed by a long, straight fracture on either side of which the pyroxene granulite has become darkened. The dark zone consists of eclogite that was formed much later, during an orogeny about 400Myr ago. Clearly the pyroxene granulite gneiss must have been deeply buried and subjected to eclogite-facies conditions 400Myr ago, but outside the narrow strip of eclogite it remained unchanged and metastable. To account for the limited extent of the alteration to eclogite, it seems that water-rich fluid travelled along the fracture and soaked a short distance into the hot, metastable gneiss on either side, where it gave that metaphorical 'push' to the perched boulder and triggered grain growth producing, in this case, the new minerals that make eclogite. The term **neocrystallization** is sometimes used to distinguish growth of new kinds of mineral grain, like this, from the simple case of recrystallization, where existing mineral grains grow. In both cases, the presence of fluid is widely thought to be essential for the process to happen.

Figure 3.20 A narrow band of eclogite (dark) with a long, straight fracture along its centre, in pyroxene granulite gneiss near Bergen in western Norway. It appears that fluid entered the fracture and soaked into the rock, where it promoted grain growth, with eclogite replacing pyroxene granulite. *Photo courtesy of Andrew Putnis.*

3.2.5 What determines the shape of a grain?

Why are the grains of some minerals commonly polyhedral (i.e. having random polygonal outlines in thin section), while grains of other minerals are commonly euhedral? Polyhedral grains are usually of quartz, feldspar, calcite, dolomite, pyroxenes, epidote, and olivine. Euhedral grains include those with **platy** shapes like muscovite, biotite and chlorite, and those with **prismatic** shapes like amphiboles, staurolite, and the Al_2SiO_5 polymorphs.

Taking polyhedral grains first, the reason for the shape is simply that it provides the lowest surface area for a given volume of mineral. A polyhedral grain with flat boundaries has less surface energy than a grain of the same mineral with the same volume, but with an irregular shape and wiggly or irregular boundaries. A polyhedron is simply a stable (minimum energy) shape. It is, incidentally, the shape adopted by bubbles in foam.

For minerals with euhedral grains, the explanation is a little different. Euhedral grains, such as plates of muscovite or prisms of andalusite, have flat surfaces that are parallel to specific internal layers of atoms. Such surfaces here provide a very tidy, and therefore energy-efficient, arrangement of the layer of atoms on the outer surface. The euhedral grain in these cases has a lower overall surface energy than a polyhedron with the same volume, despite its larger surface area.

3.2.6 Why do some minerals occur as porphyroblasts?

The small number and large size of porphyroblasts can be traced back to a process called **nucleation**. A grain of garnet can only grow where atoms can add to an *existing* grain of garnet. A metapelite in the biotite zone contains absolutely no garnet. As the temperature rises and the metapelite finds itself in the garnet zone, garnet would like to grow, but with no existing garnet grains it cannot grow. New tiny grains of garnet, called nuclei, first have to start from scratch as tiny seed crystals. The formation of a crystal nucleus is known as nucleation. Nucleation of garnet can only happen when the concentration of garnet, dissolved in the intergranular fluid, is enormous – when the fluid is said to be highly **supersaturated** or to have a very large excess of Gibbs energy, G. This state is only reached when the temperature has gone some way beyond the temperature where garnet first becomes stable (Fig. 3.21).

Once a garnet nucleus has formed, the situation changes. The nucleus grows rapidly, fed by the atoms dissolved in the

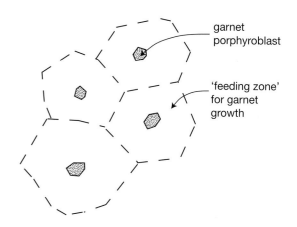

Figure 3.22 Garnet porphyroblasts. They grow large because further crystals of garnet will not nucleate in a wide 'feeding' zone around each porphyroblast where the level of excess Gibbs energy (the level of supersaturation in the intergranular fluid) is kept below the critical level needed for nucleation.

highly supersaturated fluid around it. As it grows, the level of supersaturation in its catchment volume of fluid, or feeding zone, drops below the high level needed for garnet nucleation. This prevents any further nuclei of garnet forming there (Fig. 3.22). So the few nuclei that do form grow, in the absence of competition, into very large crystals. It is a case of the 'rich getting richer'.

How does a garnet porphyroblast make space for itself as it grows? There is little evidence to suggest that it forcefully heaves aside the surrounding grains. Instead, it seems that the surrounding grains obligingly dissolve in the grain-boundary fluid just as fast as new garnet is precipitated from that fluid onto the advancing garnet crystal surface.

Why do some garnet porphyroblasts contain inclusion trails of other minerals? In some cases, the grains surrounding a growing porphyroblast, particularly grains of quartz, fail to dissolve completely, and what remains of them becomes engulfed by the porphyroblast, where they are now seen as inclusions. Very rapid growth of porphyroblasts, which can happen in the period immediately following nucleation when the level of supersaturation is still close to its peak, can leave the porphyroblasts riddled with inclusions, i.e. leave them as poikiloblasts. A case in point is the formation of poikiloblasts of cordierite in andalusite-cordierite hornfels (Fig. 2.31).

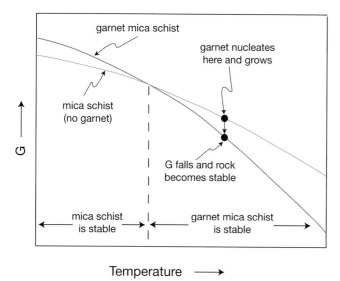

Figure 3.21 Conceptual diagram showing Gibbs energy for mica schist with no garnet (green line), which is lowest and most stable in the biotite zone, and for garnet-mica schist (red line), which is most stable in the garnet zone. Garnet can only nucleate when, with increasing temperature, the gap between the two lines becomes sufficiently large.

3.2.7 How does directed stress cause a foliated texture?

Directed stress leads to strain. When a body of rock is being strained it becomes flattened in one direction and extended in another, and platy or prismatic grains for some reason become aligned with the direction of extension (which is generally at a high angle to the applied stress) to give slaty cleavage or schistosity.

A possible mechanism for producing schistosity is grain growth under directed stress. Mica crystals have a platy shape because atoms add on to their edges much more readily than to their flat top and bottom surfaces. If one starts with a fine-grained metapelite in which tiny plates of mica are all *randomly* orientated then, as the grains grow under directed stress, those that are 'flat-on' to the stress will grow easily, allowing the rock to accommodate the strain; those that are 'edge on' will not grow so easily and, by default, will shrink and disappear, leading eventually to a rock where all the 'successful' mica plates are 'flat-on' to the directed stress (Fig. 3.23).

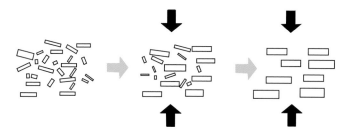

Figure 3.23 Cartoon illustrating three progressive stages in the development of schistosity as a result of grain growth with directed stress (black arrows). See the text for an explanation.

All grains, even polygonal grains, can become elongated while growing under the influence of directed stress, because they grow more easily in the direction that accommodates the strain. This will lead to a general elongation of grain shapes in the direction of the extension. An example is seen in the elongation of polygonal olivine grains in the dunite in Figure 2.71.

3.2.8 How does mylonite differ from cataclasite?

Cataclastic rocks, which include fault breccia and fault gouge (see chapter 1), are products of fracturing and comminution of the protolith. Mylonite has historically been classified as a cataclastic rock, but mylonite is now regarded as having a subtly different origin from cataclasite. Mylonite formation evidently does not involve the fracturing of grains; instead it is a process that reduces grain size by introducing enormous energy into grains as so-called elastic strain, which is then released by the spontaneous nucleation and growth of numerous tiny grains that are not strained.

In view of their tiny grain size, cataclasites and mylonites both have extreme surface energy, so they will increase their stability through grain growth at the slightest opportunity, e.g. through the introduction of fluid or through increased temperature. Following even a little grain growth, mylonite and cataclasite become difficult to tell apart. Recrystallized mylonite, known as blastomylonite, is shown in Figure 2.69 (Darwin's augen gneiss) and another example appears later, in Figure 5.3. An example of a possible recrystallized cataclasite is shown in Figure 3.24. Here, the original rock was made entirely of coarse grains of clinopyroxene. The strain suffered by its porphyroclasts is spectacular, but the matrix, seen at a high magnification, has well-formed, if small, polygonal grains of recrystallized clinopyroxene.

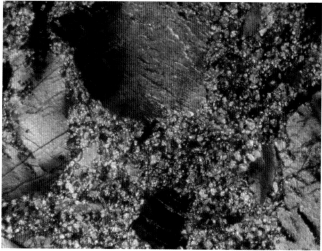

Figure 3.24 Thin section in XP of a partly recrystallized cataclasite formed by the crushing of a rock composed entirely of clinopyroxene. The porphyroclasts of clinopyroxene are so deformed that the interference colour varies from place to place within them. The tiny grains between the porphyroclasts have polygonal shapes and so have recrystallized. The field is about 5mm wide.

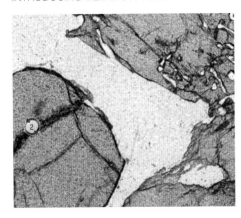
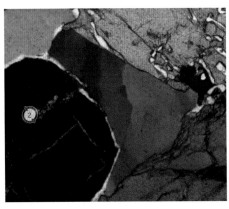

Figure 3.25 Screen shots of VM GeoLab eclogite, M26, at rotation 2 in PPL (left) and XP showing strain shadows in quartz (centre) with garnet (left) and omphacite (top and bottom right). Width of field is 0.8mm.

A widespread example of strain followed by recrystallization is seen in the familiar **strain shadowing**, or **undulose extinction**, of quartz, where the new sub-grains, misorientated by only one or two degrees, can become quite large, and are separated by sharp internal boundaries (well displayed online at rotation 2 of GeoLab M26) (Fig. 3.25).

Another question about quartz is how it develops as quartz ribbons in some blastomylonites (see Fig 5.3). Here it seems that the ribbons are continuous single crystals that have replaced finely comminuted or highly strained and streaked out quartz, with enormous surface energy, through complete and perfect recrystallization starting from a single quartz nucleus. In terms of grain growth, this is a case of 'winner takes all'.

4   Aureoles, orogenies and impacts

This chapter revisits three of the geological settings for metamorphism that were introduced in chapter 1. It describes further kinds of contact metamorphic rocks and discusses their origins. It looks in more detail at how pressure and temperature change through time in evolving mountain belts, and considers other metamorphic processes at convergent plate boundaries, particularly in rocks that undergo subduction. Thirdly, it reviews the effects of shock metamorphism at the sites of ancient giant meteorite impacts on the Earth and on extraterrestrial bodies.

4.1 Contact metamorphism

Metamorphism in thermal aureoles was introduced at the end of chapter 1, and a cordierite-andalusite hornfels (Fig. 2.30), and an amphibolite hornfels (Fig. 2.62), both with randomly oriented grains reflecting a stress-free environment, were described in chapter 2. Here, several more kinds of contact metamorphic rock have been selected for description and comment. The choice of examples is not comprehensive, but illustrates something of the variety of contact metamorphic rocks and their different origins.

4.1.1 The pyroxene hornfels facies

Just over a century ago the Norwegian geochemist Victor M. Goldschmidt, in a pioneering study of low-pressure contact metamorphic rocks around granite intrusions in the Oslo region, was able to demonstrate that, under a given set of pressure and temperature conditions, a rock's mineral assemblage varies systematically and in harmony with its overall chemical composition. Specifically, Goldschmidt recognized a consistent link between the mineral assemblages of different kinds of hornfels and the compositions of their protoliths. All the rocks were sampled from very close to granite contacts, and Goldschmidt reasonably assumed that the P–T conditions were about the same for each of them. He summarized his observations graphically (Fig. 4.1) by devising a triangular representation for the mineral assemblages, which is now familiar as the 'ACF' triangle. Goldschmidt's triangle had the oxides, Al_2O_3, CaO and combined ($FeO + MgO$) at its

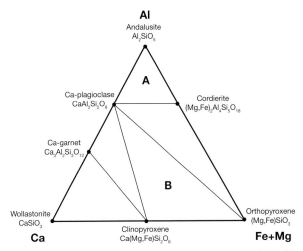

Figure 4.1 Triangular diagram with five internal triangles whose corners show coexisting minerals found by V.M. Goldschmidt in contact metamorphic rocks with different compositions right next to granite bodies in the Oslo region of Norway. It is shown with elements at its corners, and not oxides as Goldschmidt used, to be consistent with the style of ACF diagram used in chapter 2. The positions marked A and B are discussed in the text.

corners but in Figure 4.1 the corners correspond to atoms (Al, Ca and (Mg+Fe)), not oxides, in order to be consistent with the format of the ACF triangle presented in chapter 2.

Goldschmidt observed seven minerals, all positioned on the edge of the triangle, three at the corners and four on the sides. Those on the sides are joined by four lines to give five small triangles. Goldschmidt noted that an individual rock consists of the minerals at the corners of a small triangle, or at the ends of a line, depending on where its composition lies. In all cases, quartz and K-feldspar may be present too, so that any rock has at most five different minerals. For example, a metapelite whose overall chemistry plots at position 'A' in Figure 4.1 consists of the minerals Ca-plagioclase, andalusite and cordierite, with or without quartz and K-feldspar, while a metabasite falling at position 'B' consists of Ca-plagioclase, orthopyroxene and clinopyroxene, with or without quartz and K-feldspar.

Goldschmidt's work was ground-breaking. It paved the way to developing the concept of metamorphic facies, which became established later by the Finnish petrologist, Pennti Eskola, as suites of rocks from other places, each presumed to have been formed under a limited range of pressure and temperature conditions, were similarly found to display a systematic relationship between the bulk rock composition and the mineral assemblage.

The minerals for position 'B' in Figure 4.1 are those seen in metabasites in the pyroxene granulite facies, showing that the Oslo rocks belong to the pyroxene granulite facies. However, in a contact metamorphic setting, the pyroxene granulite facies has historically been named the pyroxene hornfels facies. The amphibolite facies, similarly, has been called the hornblende hornfels facies.

Two of the minerals named in Figure 4.1 are less well known. Both are calcium-rich and occur in impure metamorphosed limestones. One is calcium garnet, also called grossular, which has the formula $Ca_3Al_2Si_3O_{12}$ and was mentioned in chapter 3 (see also Appendix 2). The other is new here. This is **wollastonite**, formula $CaSiO_3$. It is a white prismatic mineral that results from a simple reaction between calcite and quartz, such that calcium carbonate plus silicon dioxide becomes calcium silicate plus carbon dioxide:

$$CaCO_3 + SiO_2 = CaSiO_3 + CO_2$$

calcite   quartz   wollastonite carbon dioxide

This reaction occurs at rather high temperatures and low pressures (Fig. 4.2) so wollastonite is limited to contact aureoles, and is not found in regional metamorphic rocks.

4.1.2 Marbles in contact aureoles

As well as grossular and wollastonite, some other minerals in contact metamorphosed limestones are worth noting. When dolomitic limestone is metamorphosed at low pressure in a contact aureole, the dolomite can break down, not by reacting with other minerals but simply by spontaneously changing within itself to a mixture of calcite and a new mineral, periclase (magnesium oxide), while carbon dioxide gas is released. The reaction is written as follows:

$$CaMg(CO_3)_2 = CaCO_3 + MgO + CO_2$$

dolomite   calcite   periclase carbon dioxide

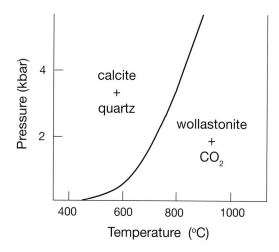

Figure 4.2 Experimentally determined stability of calcite plus quartz. The reaction between them to give wollastonite + CO_2 occurs naturally in contact metamorphic settings where the temperature is high and the pressure is very low.

This is called de-dolomitization, and a famous example of it occurs in the aureole of the granite at Beinn an Dubhaich on the island of Skye in NW Scotland. The periclase grew as numerous tiny cubes inside grains of dolomite, as the dolomite changed to calcite (Fig. 4.3). However, the periclase did not survive. It reacted with water at some later stage when water infiltrated the marble and hydrated it into another new mineral, brucite (magnesium hydroxide). The reaction in this case is:

$$MgO + H_2O = Mg(OH)_2$$

periclase   water   brucite

During the hydration process, each cube-shaped crystal of periclase retained its approximate shape. Brucite is a colourless, low-relief mineral with grey interference colours; it appears as random flakes filling the 'cubes' of former periclase. Each cube-shaped aggregate of brucite is thus a **pseudomorph** of brucite after periclase. The rock can be found in the UK Virtual Microscope collection, named 'brucite marble'.

As with the formation of wollastonite, evidence of de-dolomitization is only seen in contact metamorphosed limestone where pressures were low and temperatures high. Under normal conditions of regional metamorphism, with a pressure of several kilobars, dolomite remains stable, even at the high

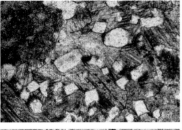

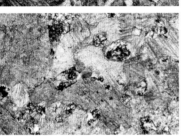

Figure 4.3 Thin section of de-dolomitized marble (brucite marble) from the island of Skye, in PPL, and (below) in XP. The width of the field is 1.5mm. Most of the rock consists of slightly brownish calcite, which appears cream in XP. The colourless squares in PPL are pseudomorphs of brucite, Mg(OH)$_2$, from the hydration of earlier periclase, MgO.

out from the granite, one finds the minerals forsterite and diopside (the Mg end-members of olivine and clinopyroxene, respectively). These two were described in chapter 2, from the forsterite marble at Glenelg (Fig. 2.38). Further out again, the mineral tremolite (pure magnesium clinoamphibole, formula Ca$_2$Mg$_5$Si$_8$O$_{22}$(OH)$_2$) is present. An account of the formation of all these minerals at Beinn an Dubhaich was published by C.E. Tilley, Professor of Mineralogy and Petrology at Cambridge University in the middle of the last century. Tilley inferred a sequence of metamorphic reactions, with increasing temperature, in which the minerals appeared in the order: tremolite, forsterite, diopside, periclase, wollastonite. Keen that his students would recall and regurgitate his observations in their examination answers, Tilley included in his publication a mnemonic to aid their memories. It was: 'Tremble! For Dire Peril Walks'.

The compositions of the five minerals concerned, along with those of calcite, dolomite and quartz, are shown on a triangle (Fig. 4.4) with the elements Ca, Mg, and Si at its corners. Six smaller versions of this triangle show how the mineral assemblage at Beinn an Dubhaich changes as the granite contact is approached from the cold country rock. The first small triangle starts with the unmetamorphosed siliceous dolomitic limestone protolith with quartz + calcite + dolomite. In each small triangle the composition of the protolith is marked by a blue spot.

71

temperature of the granulite facies. The very low pressures of contact metamorphism allow fluids (H$_2$O as well as CO$_2$) to be lost from minerals at temperatures perhaps 200°C lower than for the same reactions in a regional setting (see Fig. 4.2), yielding, at these quite modest temperatures, mineral assemblages that would normally be associated with high-grade metamorphism.

The evidence for de-dolomitization and the formation of wollastonite at Beinn an Dubhaich is found close to the granite contact, where temperatures were highest. Further

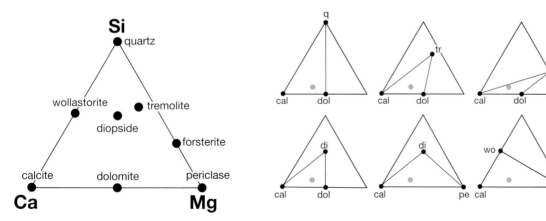

Figure 4.4 Triangular diagram showing the positions of minerals in siliceous dolomitic marble in terms of the elements Si, Ca and Mg, with six smaller triangles showing assemblages of three minerals that occur in the aureole, progressively closer to the contact with the Beinn an Dubhaich granite. The blue spot is the composition of the whole rock.

4.1.3 Metapelites in contact aureoles

Thermally metamorphosed pelitic rocks display many features in addition to those described in chapter 2. Four examples chosen here for description and comment are an occurrence of sapphire, a case of andalusite with a multistage history, aureoles with sillimanite instead of andalusite, and aureoles with pelitic schist instead of hornfels.

Sapphire is the blue variety of corundum, another new mineral here. Its chemical composition is Al_2O_3. An example of its occurrence is in the aureole of a diorite pluton near the town of Comrie, just north of the Highland Boundary Fault in Scotland. The protolith is low-grade pelitic schist of the Grampian belt. In PPL the corundum is patchily blue in colour, so it can just about be called sapphire (Fig. 4.5). However, it is far too small and dusty to be considered of gem quality. Andalusite occurs with the corundum in this hornfels (though it is not seen in Fig. 4.5).

Corundum is quite rare in metapelites, despite its Al-rich composition. The reason for this is that metapelites almost always contain quartz, and quartz and corundum cannot coexist. The two would react to make kyanite, sillimanite or andalusite. In this rock it seems that quartz was in short supply, and that after all available quartz had been used to make andalusite, some Al_2O_3 was left over to form the corundum. Rocks having insufficient SiO_2 to allow free quartz to appear are described as silica-deficient. The protolith here was presumably a silica-poor shale, dominated by Al-rich clay minerals. This suggestion is quite plausible, bearing in mind that a hypothetical metapelite consisting entirely of muscovite would become converted on dehydration to an assemblage of K-feldspar and corundum (no quartz) by the reaction:

$$KAl_2Si_3AlO_{10}(OH)_2 = KAlSi_3O_8 + Al_2O_3 + H_2O$$

muscovite   K-feldspsar   corundum   water

The second example of baked metapelite contains andalusite with a multistage history. The rock was collected from beside the Ardara granite in NW Ireland, which again was intruded into Grampian low-grade schist. The rock is available online in the VM GeoLab collection as M20 andalusite hornfels, and a screenshot is shown in Figure 4.6.

The andalusite occurs as randomly oriented prismatic porphyroblasts. Cross-sections of them appear as squares, diamonds, parallelograms, and rectangles, depending on the

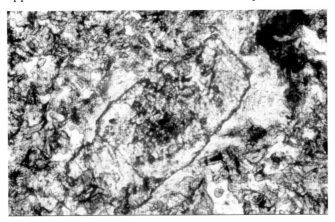

Figure 4.5 Thin section in PPL of a silica-deficient pelitic hornfels with a large sub-rectangular grain of corundum, Al_2O_3. Its mottled blue colour means that technically it can be called sapphire. It is about 0.2mm long.

Figure 4.6 Andalusite hornfels – M20 in PPL and in XP. For a description see the text.

angle they make with the thin section. Many of them show the cross-shaped pattern of tiny inclusions characteristic of andalusite and mentioned in chapter 2. The surrounding rock is fine-grained, typical of hornfels, and contains brown pleochroic biotite, some small plates of muscovite (showing bright interference colours), and quartz.

The andalusite appears to have grown in two stages. Pale pink, pleochroic, inclusion-free cores with colourless overgrowths are visible in some porphyroblasts (Fig. 4.6, and online at rotations 1 and 2). The pink cores have scalloped edges suggesting that they had begun to dissolve before the second phase of andalusite grew on top of them. Numerous inclusions in the colourless andalusite suggest an initial burst of rapid growth, capturing existing fine-grained minerals too slow to dissolve up and escape the advancing andalusite. The outermost andalusite is inclusion-free (except for the cross-shaped diagonals). The reason for two stages of growth is not clear. It may be related to the intrusion history of the granite, which is reported as having arrived in two phases, and may therefore have led to two separate pulses of heating. The reason why the andalusite cores are pink and pleochroic is not understood. The pleochroic and non-pleochroic varieties have been analysed and found to have almost identical chemical compositions.

A later stage in the rock's history is retrograde metamorphism around andalusite grain margins. Two styles of alteration are seen, both yielding muscovite. In one, aggregates of very fine-grained muscovite have 'eaten away' at the margins. In the other, well-formed and randomly orientated single plates of muscovite have partly replaced the andalusite, and simultaneously grown in the main body of the rock. The formation of muscovite at the expense of andalusite requires the input of water. The source of the water is not known, but three possibilities can be suggested. Firstly, as the granite and its aureole cooled, thermal convection of groundwater may have been established, as envisaged for hydrothermal metamorphism (Fig. 1.10). Secondly, water may have been dissolved in the magma and released into the country rock during the final stages of granite crystallization. Thirdly, water movement may be linked to the metamorphism itself. Schematic temperature profiles at successively later times across an igneous contact are shown in Figure 4.7. They suggest that, at times marked 4 and 5, while rocks at some distance away from the contact are still getting hotter and releasing

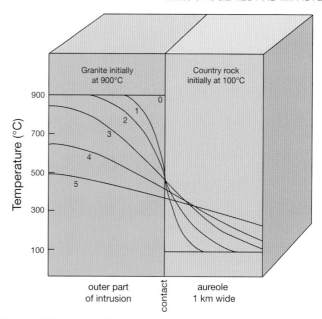

Figure 4.7 Schematic temperature profiles at six different times, starting from the moment of intrusion, across the contact between a body of granite at 900°C and country rock at 100°C. Note that the inner aureole from time 3 onwards is already cooling down (and becoming water-deficient) while the outer aureole is still getting hotter (and releasing water).

water, rocks close to the contact are already cooling, so will be water-deficient (see chapter 3) and capable of absorbing the water released further out.

The third example of baked metapelite concerns the presence, or not, of sillimanite in an aureole, and its bearing on the depth of metamorphism. Three different situations have been observed in aureoles. In the first, sillimanite extends throughout the aureole except perhaps on the cool outer fringes, where andalusite may occur. In the second, sillimanite is present close to the contact while andalusite occurs further out. In the third, andalusite is present throughout the aureole.

The Al_2SiO_5 P–T diagram, introduced in chapter 2 (Fig. 2.32), suggests that these three situations correspond, respectively, to deep, intermediate, and shallow levels beneath the surface (Fig. 4.8). In all three cases it is assumed that the highest temperature reached in the aureole, i.e. at the contact, is about 700°C, and that andalusite will only start to grow if the temperature exceeds about 500°C.

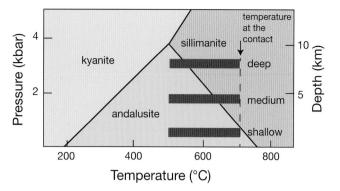

Figure 4.8 The pressure–temperature diagram for Al_2SiO_5 (kyanite, andalusite and sillimanite) showing three bands in red corresponding to P–T conditions in a deep aureole (with sillimanite, except at the cool outer margin), an aureole at medium depth (andalusite in the outer aureole, sillimanite close to the contact), and a shallow aureole (andalusite throughout).

Some granite bodies, incidentally, have no contact aureole. A case in point is the Oughterard granite near Galway in the west of Ireland, which is again located within rocks of the Grampian orogenic belt. The absence of an aureole was puzzling until the age of the granite was measured. It was found to be as old as the Grampian orogeny (about 470Myr), so the country rocks were presumably still hot following the orogeny when the granite invaded.

The fourth and final example concerns aureoles in which pelitic rocks occur as schist rather than hornfels. An example is the aureole of the Leinster granite **batholith** in south-east Ireland. Here andalusite is developed as usual, but the matchstick-like prisms tend to lie in a plane parallel to the schistosity which, in its turn, is parallel to the granite margin. Schistosity is produced when grains grow under the influence of directed stress. Normally when granite is being intruded, the magma flows into a space that opens up for it. No directed stress is involved and hornfels is the metamorphic product. In some cases, however, the intrusion mechanism is described as 'forceful'; it is as though the magma has forced its way in, making space for itself by heaving the country rock aside. Directed stress during cooling of an aureole may also arise where magma is intruded into country rocks that are, themselves, actively being deformed under directed stress. This was probably the situation with the Leinster granite

Figure 4.9 St Kevin's Church in the ruined AD 600 monastic settlement at Glendalough, Co. Wicklow, Ireland. It is built from schist collected from the aureole of the Leinster granite. The roof is even made from schist. A spectacular four-hour loop walk through the aureole, into the granite, and out again is a popular visitor attraction here. *Image Wiki commons attribution is By Warrenfish – Own work, Public Domain, https://commons. wikimedia.org/w/index.php?curid=4874097.*

batholith, where schist extends over a horizontal distance of about 3km from the granite contact, and is beautifully exposed in the glacial valley of Glendalough (Fig. 4.9).

4.2 Metamorphism in orogenic belts and subduction zones

In chapter 1 it was explained that regional metamorphism results from orogenesis (mountain building) where continental crust becomes much thicker owing to lateral compression, and is then restored to its original 'stable' thickness of about 35km when the agencies of buoyancy and erosion bring the root zone of the mountains to the surface. While this general picture provides a simple working model for the origin of metamorphic belts, the details of orogenesis can be complex and varied. This section explores several of those details, including the variation of P–T conditions among different orogenic belts; the paths followed through time by suites of rock during their burial, heating, decompression and cooling; the style of metamorphism related to subduction zones; and how the timing of events in a rock's metamorphic history can be determined.

4.2.1 Low-, normal-, and high-pressure metamorphic belts

The first of the details to be explored here concerns the pattern of isograds in the Grampian orogenic belt in Scotland. The isograds mapped by George Barrow close to the Highland Boundary Fault (Fig. 2.27) have since been mapped over the entire Grampian belt in Scotland (Fig. 4.10) and their distribution is less straightforward than the discussion in chapter 2 might suggest. They were not easy to map because wide tracts of country have no metapelites suitable for identifying Barrow zones. However, mineral assemblages in calc-silicate rocks were found to change with grade and to provide a reasonable proxy for metapelites.

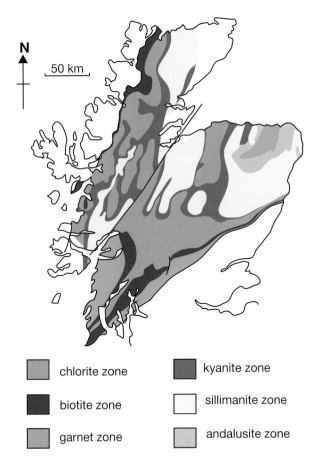

Figure 4.10 Barrow zones in the Grampian orogenic belt in Scotland. The staurolite zone is included as part of the kyanite zone. The grey colour in the east is a zone of regional andalusite.

In the south-west of Scotland, the pattern of isograds is straightforward, with the highest grade rocks (in the garnet zone) forming a strip along the centre of the belt, flanked on the NW and SE sides first by rocks of the biotite zone and, further out, by rocks of the chlorite zone. However, in the north-east of Scotland, in a region named Buchan, the pattern changes, and a large area is characterized by the presence of andalusite in metapelites (coloured grey in Fig. 4.10). This contrasts with the regular Barrow zones, where kyanite and sillimanite occur, and andalusite does not. The andalusite implies that the pressure in the Buchan region was unusually low, below 4kbar based on the Al_2SiO_5 diagram. This means that the uplift and exhumation were much less here than elsewhere. The style of metamorphism in the Buchan region has been described as low-pressure metamorphism, or Buchan-type metamorphism; in contrast, regional metamorphism to the southwest, with regular Barrow zones, is called 'normal' metamorphism, or Barrovian metamorphism.

The high temperatures and low pressures inferred for the Buchan rocks are not vastly different from those in some contact aureoles, and gabbro plutons in the Buchan region, which were intruded *during* the Grampian orogeny, have been implicated as the source of the high temperatures. It is imagined that their aureoles simply merged into one continuous realm of high temperatures.

Turning to other metamorphic belts of the world, some display evidence of a very different suite of metamorphic P–T conditions, neither Buchan nor Barrovian in style. These belts are characterized by abnormally high pressures and rather low temperatures. They are recognized because metabasites within them are blueschist or even eclogite. Such metamorphism has been named high-pressure metamorphism.

The approximate P–T conditions inferred for these three regimes of regional metamorphism – the low-pressure (Buchan), the normal (Barrovian), and the high-pressure regimes – are shown in Figure 4.11 as broad arrows superimposed on the metamorphic facies diagram from Figure 2.61.

How are these three different regimes of orogenic metamorphism explained? The beginnings of an answer emerged more than 50 years ago when a Japanese geologist named Akiho Miyashiro noticed that regional metamorphic rocks along the southern margin of the main island of Japan define two parallel, but contrasting, belts of the same age (Fig. 4.12). To the south, the so-called Sanbagawa belt is characterized

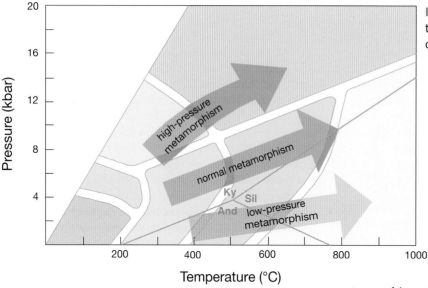

Figure 4.11 Metamorphic facies diagram with three broad arrows each corresponding to a distinct regime of regional metamorphism.

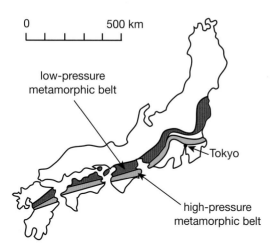

Figure 4.12 Simplified map showing the so-called paired metamorphic belts of the Japanese islands, recognized by Ahiko Miyashiro.

by metamorphic rocks that were formed at a high pressure and low temperature, and include blueschist-facies rocks. The protoliths in this belt are mostly sedimentary rocks and basic igneous rocks of the oceanic crust. North of the Sanbagawa belt is the Ryoke belt, which is tentatively linked to the Abukuma plateau further east. The Ryoke–Abukuma

metamorphic rocks are andalusite-bearing, and are invaded by numerous igneous intrusions, like those in the Buchan district of Scotland.

Miyashiro dubbed the two adjacent strips 'paired metamorphic belts' and was years ahead of his time when he proposed that the juxtaposition of a belt of cold, deeply metamorphosed rocks and a belt of hot, shallowly metamorphosed rocks might be an expression of the kind of geological activity in present-day Japan, where a deep-ocean trench runs parallel to a chain of on-shore volcanoes. Plate-tectonic theory had not been developed at that time, but the rudiments of subduction were beginning to be revealed by earthquake foci located on a plane that sloped downwards from the trench beneath the volcanoes. Miyashiro realized that the volcanoes were the surface expression of abundant igneous intrusions, and a correspondingly hot environment, at a shallow depth, and that such a setting would be suitable for a low-pressure, andalusite-bearing metamorphic belt. He also realized that an oceanic trench is a place where cold sediment, including a huge amount of detritus from the volcanoes, would form a thick accumulation, which would conceivably be thickened further by whatever 'down-dragging' process was generating the earthquakes. Thus, a trench setting seemed to be suitable for the kind of high-pressure metamorphism inferred for the Sanbagawa belt. Miyashiro's vision is illustrated in Figure 4.13, a hypothetical cross-section through an active trench and

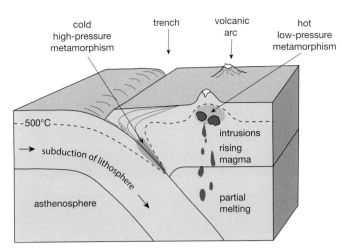

cold
high-pressure
metamorphism

trench

volcanic
arc

hot
low-pressure
metamorphism

-500°C

subduction of lithosphere

asthenosphere

intrusions

rising
magma

partial
melting

Figure 4.13 A hypothetical cross-section through an ocean trench and volcanic arc showing how paired high-pressure and low-pressure belts may be related to subduction, as envisaged by Ahiko Miyashiro.

its parallel volcanic chain. This vision helped at that time to revive a latent curiosity about the 'big picture' among geologists who, previously, had studied metamorphic rocks in isolation and had not seriously concerned themselves with how they might relate to large-scale geological processes.

With the full exposition of plate-tectonic theory in the early 1970s, rapid progress was made at the intersection of petrology and movements of the newly recognized lithosphere. The simple concept of lateral compression and thickening of the crust portrayed in chapter 1 was replaced by the more elaborate idea that continental crust is thickened through plate subduction. This led to new ways of interpreting the wealth of information locked up in the textures and mineral assemblages of regional metamorphic rocks.

4.2.2 Subsurface temperatures and P–T–t paths

One of the developments since the 1970s is a better understanding of the temperature beneath the surface. Heat in the Earth is continually being generated by the slow decay of **radioactive** uranium (U), thorium (Th) and potassium (K). These three elements are many times more concentrated in rocks of the continental crust, compared with the mantle, because igneous processes over billions of years have scavenged them from the mantle and brought them into the crust. The heat

produced by their decay is called **radiogenic heat**, and it is lost through the Earth's surface. The rate of loss averages about one-tenth of a watt per square metre – roughly the same as a 1kW room heater for the area of a large soccer field. This is, incidentally, less than 0.1% of the rate that radiant energy reaches the Earth from the Sun, yet it is still more than enough to drive plate movements, fuel volcanoes, generate earthquakes, and induce metamorphism.

In continental regions it is estimated that about half of the heat escaping through the surface comes from the U–Th–K enriched rocks in the crust, and that half comes from the underlying mantle. The heat moves through the crust by a process called **conduction**, which causes the temperature to increase with depth. The rate of increase is called the temperature gradient or **geothermal gradient** (see Appendix 1), and it tends towards a 'steady state' where the heat being supplied from below equals the heat being lost through the surface. A typical geothermal gradient in the upper crust is 20°C per km. If the heat flow were higher, the temperature gradient would be correspondingly higher, in proportion. So if the radiogenic heat production in the continental crust were to double, the total amount of heat (now two-thirds from the crust and one-third from the mantle) flowing through the surface would increase by 50%, and the temperature gradient would therefore also rise by 50% to 30°C per km. A doubling of heat production in the crust would happen, in theory, if the thickness of the crust were to double from 35km to 70km during orogenesis. However, it would take a long time for the steady gradient of 30°C per km to become established in this newly thickened crust. The initial thickening of the continental crust would actually *reduce* the average temperature gradient, in the short term, to *less* than 20°C per km, and perhaps as little as 10°C per km, as cool rocks from near the surface are deeply buried, but radiogenic heating would slowly warm these rocks, and, given enough time, eventually establish a steady state, with a gradient of 30°C per km. However, time is short; continental crust that is 70km thick is buoyant and will lose pressure rapidly through erosion at the same time as it is warming up, and this complicates matters.

The interplay of burial, radiogenic heating and loss of pressure is thought to lead to a situation where a part of the metamorphosed crust that ends up exposed at the surface will have followed a changing pressure–temperature path over time (a **P–T–t path**) like the one shown in Figure 4.14. The

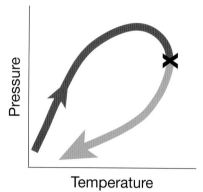

Figure 4.14 A postulated P–T–t path followed by rocks in the continental crust during orogenesis. For an explanation see the text.

Figure 4.15 Screen shots in PPL (upper image) and XP of VM GeoLab Glaucophane Schist, M13, from Achill, western Ireland. The uniform grey area on the right in XP is a single large poikiloblast of Na-plagioclase. For comment see the text. Width of field is 2.5mm.

volume of rock, with burial and thickening of the crust, will first have increased in pressure faster than in temperature, and will then have warmed up due to radiogenic heating, continuing to get hotter even as it was passing through its greatest depth. It will have reached its peak temperature while losing pressure during its early period of decompression, and then will have cooled down as it continued to lose pressure and approached the surface.

The P–T–t path in Figure 4.14 is shown in red where the temperature is increasing, and in blue where the temperature is dropping. The colours are chosen as a reminder that during heating (the prograde stage), water is being continually released, which keeps the rocks constantly close to a state of equilibrium. During cooling (the retrograde stage, coloured blue), the rocks will most likely be water-deficient, and the state of equilibrium in minerals and texture achieved at the highest temperature is likely to be preserved, in a metastable state, as was explained in chapter 3 (section 3.1.6).

Three lines of evidence suggest that rocks from the Grampian orogenic belt followed a P–T–t path like the loop shown in Figure 4.14.

Firstly, rare blueschists are known from near the southern margin of the Grampian belt on the island of Achill in western Ireland, where they occur as local pockets within a greenschist-facies region. Their textures are consistent with the rocks having been metamorphosed in the blueschist field during early subduction-related burial when they were cold,

and then later having been warmed up and decompressed, and turned into greenschist facies rocks. For example, Figure 4.15 shows glaucophane apparently trapped as metastable inclusions inside large poikiloblasts (inclusion-filled grains) of albite that may have grown as the blueschists warmed up and lost pressure.

Independent evidence that blueschist-facies material was being eroded from the Grampian mountain belt, before having time to warm up, is preserved in occasional sand grains made of glaucophane in a thick unit of sandstone that was deposited on the floor of an Ordovician ocean to the south of the orogenic belt.

Secondly, the peak temperature in the Grampian orogeny was probably reached after deformation was completed, based on the late growth of the index mineral garnet in the garnet-mica schist, M08 (Fig. 2.21), and the late growth of the index mineral staurolite in the garnet-mica schist, M10 (Fig.

2.25), both described in chapter 2. Presumably these rocks were deformed strongly during the period of subduction and thickening of the crust, and by the time of the peak temperature during later uplift and exhumation, driven by buoyancy, directed stress had ceased. A similar case can be made for the epidote amphibolite, M02 (Fig. 2.48) where the sodic plagioclase protrudes a short distance along the boundaries between surrounding grains of epidote and hornblende, so presumably grew later. Since the plagioclase has a low density, its formation is plausibly a response to decompression while the temperature was still rising.

Thirdly, evidence that the peak temperature was reached after regional deformation was over can be inferred from the pattern of Barrovian isograds, particularly in the south-west of Scotland where the pattern is simple (Fig. 4.10). So-called 'way-up' structures, like cross-bedding (see Fig. 2.4) in the metasediments here, testify to complex, large-scale folding and faulting, with huge parts of the crust having been turned completely upside down. If the peak temperature conditions had been reached *before* this major deformation, the isograds would have been similarly folded and deformed, but they have hardly been affected by deformation, so must have become established later, when directed stress was no longer a factor.

Changing the discussion, but staying for a moment with the isograds in Figure 4.10, one might imagine that they record temperature–depth profiles, or fossil geotherms, for different places in the orogenic belt at the peak of metamorphism. However, they probably do not. They probably record peak temperature conditions at progressively later times going from the margins to the centre of the belt. A hypothetical set of nested loops is shown in Figure 4.16, with the smallest loop representing a place near the margin of the belt, and the largest representing a place near the centre. The filled black circle on each loop marks its peak temperature. Assuming that all the rocks were losing pressure in tandem, the low-grade rocks, with less distance to rise, could well have been back at the surface and cold while the high-grade rocks were still getting hotter and just reaching their peak temperature. The dashed line through the circles is not a fossil geotherm, but a trace of P–T conditions at peak temperatures at successively later times.

A related point of interest is that the exhumation and cooling part of a P–T–t path may not always be caused simply by buoyancy and erosion. In recent years it has been

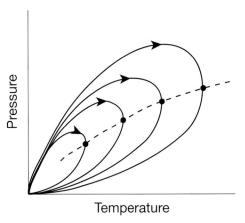

Figure 4.16 Four hypothetical pressure–temperature paths followed by four separate packages of rock at different places in an evolving orogenic belt. See the text for an explanation of why the dashed line joining the black filled circles (the peak temperature for each package) is not a fossil thermal gradient.

recognized that high-grade metamorphic rocks in some young orogenic belts have locally been brought rapidly to the surface apparently because the crust was *stretched*. This is the case, for example, for the migmatite from the island of Naxos in Greece, decorating the cover of this book. The high-grade rocks occupy a central dome, called a **metamorphic core complex**, separated from overlying low-grade or unmetamorphosed sediments by a gently sloping mylonite shear zone, thought to have been produced during extension of the crust. The origin of core complexes is not fully understood; as noted in Appendix 1, extension of the continental lithosphere normally causes subsidence and sedimentary basins, and not features like core complexes with positive relief.

4.2.3 Measuring little 't' in a P–T–t path
The science of obtaining quantitative ages from minerals and rocks is known as **geochronology**. Geochronology puts dates on *events* in a rock's history, such as the time in the past when magma crystallized, or the time when a sediment was lithified, or the time when a metamorphic rock cooled below a particular temperature. These 'ages' are obtained by a method known as **isotopic dating**, and they help to constrain little 't' in a rock's P–T–t path. To give some idea of how isotopic dating works, the results from two separate studies relating to the

Grampian orogenic belt are chosen. These studies make use, respectively, of the uranium–lead (U–Pb) dating of zircon and the potassium–argon (K–Ar) dating of biotite. The principles of these two methods are explained in Appendix 5.

The results for the Grampian belt are as follows. U–Pb dating of zircon generally yields the time before the present when a zircon crystal grew. If it grew in an igneous rock, its age is also the age of **crystallization** of the magma. Regarding the Grampian orogeny, the method has confirmed that the main orogenic event happened at around 470 million years (Myr) ago. The zircon in question grew in a body of gabbro that was intruded *during* the Grampian orogeny. The gabbro cuts across an earlier Grampian foliation in the surrounding schists, but has itself been deformed by a later Grampian phase of directed stress.

The K–Ar dating of metamorphic biotite dates a different 'event' in the evolution of an orogenic belt. During the early days of isotopic dating, many samples of mica schist from the Grampian orogenic belt were collected and dated by the K–Ar method, and many biotite ages were published. They were found to vary widely, but to correlate quite well with the location of the sample in the belt (Fig. 4.17). Low-grade rocks from both the NW and SE margins gave biotite ages of about 460Myr, while some rocks from the sillimanite zone in the centre produced biotite ages that were younger than 400Myr. These results raised two questions. Why do the ages vary with location, and why are they all younger than 470Myr, which is the age of the orogeny based on the U–Pb method?

The reason for the large range of ages is now clear. It turns out that a K–Ar age for biotite does not date the time when the biotite grew, but rather it gives the time after metamorphism when the biotite cooled down. Specifically, it dates the time when its temperature fell below about 300°C. This is the so-called **blocking temperature** for biotite, as is explained in Appendix 5. The pattern of ages shown in Figure 4.17 therefore implies that rocks in the centre of the orogenic belt were the last to cool down below 300°C, which is as one might expect, considering the amount of rock that had to be removed from above them. They took more than 70Myr to do so. The rocks near the NW and SE margins of the belt, in contrast, were not buried so deeply, and were exhumed much more quickly, within 10Myr of the time (470Myr ago) of peak metamorphism.

The pattern of biotite K–Ar ages in Figure 4.17 is broadly consistent with the nest of P–T loops in Figure 4.16. The

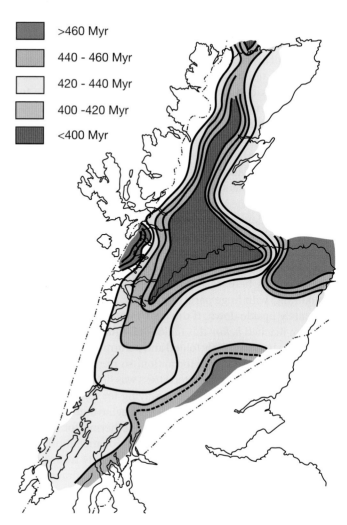

Figure 4.17 The Grampian metamorphic belt in Scotland contoured in K–Ar ages obtained from biotite samples. Note that the shape of Scotland has been changed to restore the Grampian belt to what some people think it was like at the time of the orogeny.

low-grade rocks near the margins of the Grampian belt, which were exhumed and cooled early, correspond to the smallest loop, while the highest-grade rocks in the centre correspond to the largest loop, and were still being heated while the marginal rocks were being exhumed.

Other potassium-bearing minerals have different argon blocking temperatures. That of muscovite is about 350°C and

>460 Myr

440 - 460 Myr

420 - 440 Myr

400 -420 Myr

<400 Myr

that of hornblende about 500°C. K–Ar ages for biotite, muscovite and hornblende from the same rock, therefore, fix three points on the rock's cooling path – the times when the rock cooled through 500°C, 350°C and 300°C, respectively. Moreover, the concept of blocking temperature extends to other dating methods (not described here), with each method constraining the time when a particular temperature was passed in the rock's cooling history.

4.2.4 High-pressure metamorphism and its geological consequences

While the Grampian belt preserves only rare blueschist-facies rocks, certain metamorphic belts in other parts of the world are composed largely of high-pressure rocks. These rocks are derived from various protoliths; they include blueschists, marbles and metagranites.

Blueschists in high-pressure belts are usually composed of glaucophane and epidote (Fig. 2.53), but in some cases a new mineral called **lawsonite** appears in place of epidote. Lawsonite is also a hydrous Ca-Al-rich mineral. Its formula is equivalent to anorthite (the Ca-plagioclase end-member) combined with two lots of H_2O, i.e. $CaAl_2Si_2O_8 + 2H_2O$. It is colourless and has a high density. Lawsonite and glaucophane are stable together at even higher pressures or lower temperatures than epidote and glaucophane. Lawsonite contains about 11% by weight of water, making it one of the 'wettest' minerals. At higher pressures and temperatures, blueschists lose their glaucophane, epidote and lawsonite as water is driven off, and garnet and omphacite develop in their place as the rock turns into anhydrous eclogite.

Marbles in high-pressure belts can contain the polymorph of $CaCO_3$ called aragonite, which is denser than calcite, as was mentioned in chapter 2. However, the aragonite does not survive very well during exhumation, and most aragonite marbles have changed substantially to calcite marbles by the time they reach the surface.

In metamorphosed granites and feldspathic sandstones albite can become converted to a combination of quartz and **jadeite** which, together, have a higher density than albite. Jadeite is clinopyroxene with the formula $NaAlSi_2O_6$ (see Appendix 2). The breakdown reaction with increasing pressure is:

$$NaAlSi_3O_8 \quad = \quad NaAlSi_2O_6 \quad + \quad SiO_2$$
$$\text{albite} \qquad\qquad \text{jadeite} \qquad \text{quartz}$$

The stability fields of lawsonite, of aragonite, and of jadeite plus quartz have all been determined experimentally. They imply cold, deep conditions, lying close to, or on the cold side of, a geothermal gradient of only 10°C per km (Fig. 4.18).

High-pressure metamorphic rocks pose a question. The P–T conditions for their stability are far removed from a steady-state geotherm (typically about 20°C per km), lying close to a gradient of only 10°C per km, consistent with the rapid, deep burial of cold rocks as a result of subduction. So why, once formed, do the rocks not warm up and recrystallize as greenschist or amphibolite facies rocks before getting back to the surface, as appears to have happened in the Grampian belt? That they stayed cold can only mean very rapid exhumation; they must have returned to the surface almost before they had begun to warm up. The tectonic mechanism for their speedy return is not fully understood.

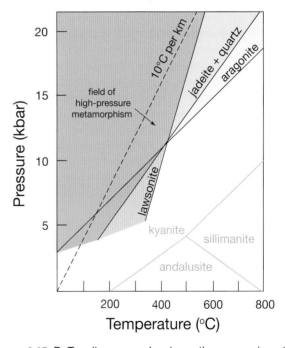

Figure 4.18 P–T diagram showing the experimentally determined stability fields of lawsonite (pink), aragonite (blue), and jadeite+quartz (grey). The approximate field of high-pressure metamorphism is indicated, where the three stability fields overlap close to the thermal gradient of 10°C/km. The Al_2SiO_5 stability fields are included for reference.

In most cases, the dense metabasites (blueschist and eclogite) made from ocean crust during its subduction probably never make it back to the surface, but continue on their downward journey into the mantle. Here, their metamorphic changes have fundamental implications for two separate plate-tectonic processes.

Firstly, the entire oceanic crust will eventually turn to eclogite. Eclogite has a significantly higher density than the surrounding mantle. Its density is typically about 3.6g/cm³, while that of mantle peridotite is closer to 3.4g/cm³. The difference gives the altered oceanic crust enormous negative buoyancy and, as it sinks ever deeper into the mantle, it is thought to pull much of the attached lithosphere down with it. Many geologists believe that the resulting down-drag, dubbed 'slab-pull', is easily the most important of the forces that cause plates to move.

Secondly, a significant fraction of the oceanic crust will probably have been hydrated to greenschist or amphibolite through hydrothermal metamorphism at the spreading ridge where it was created (see Figs 1.10 and 1.11). With subduction, metamorphic changes will first result in the water being transferred from chlorite, epidote and hornblende to different hydrous minerals such as lawsonite and glaucophane as conditions move into the blueschist facies. With further subduction (and heating), however, eclogite will eventually be formed and the water will be driven off. This water is thought to move vertically and infiltrate the hot mantle peridotite in the overlying wedge of asthenosphere above the subduction zone, and to be instrumental in triggering the major volcanism seen at the surface.

The mechanism inferred here is that the infiltrating water induces partial melting in the hot peridotite it enters, yielding a small fraction of hydrous basaltic liquid. This is because peridotite at a high pressure in the presence of water will *start* to melt at a significantly lower temperature than 'dry' peridotite, in the same way as 'wet' granite under pressure starts to melt at a much lower temperature than dry granite (see chapter 3, Fig. 3.15). However, the amount of melting will be only slight; substantial melting cannot occur until the asthenosphere rises and undergoes decompression melting, as happens beneath mid-ocean ridges. The huge scale of volcanism above subduction zones suggests that the asthenosphere must rise here too, but how it does so, where plates *converge*, is not clear. Perhaps the drag of the descending slab 'sucks'

the lithosphere on both sides of the trench towards the subduction zone, causing extension of the over-riding lithosphere and allowing the very slightly molten, and hence softened, peridotite to start 'flowing' upwards.

The presence of dissolved water in the resulting basaltic magma might explain why many of the volcanoes erupt as andesite. Water allows the hydrous mineral, hornblende, to crystallize from the magma. Since hornblende has a very low percentage of silica, if it is removed (for example, by the crystals settling out under gravity) the residual magma will be enriched in silica, giving it an intermediate, or andesitic, composition.

Finally, where the rising magma enters continental crust on the over-riding plate, as it does, for example, in Japan and along the Pacific coast of South America, the invasion by so much basic and intermediate magma would lead to very high temperatures in the continental crust, causing metamorphism, migmatite formation and the production of granitic melt from pelitic protoliths, as explained in chapter 3. Thus, volcanoes on continental crust above subduction zones are not just basaltic and andesitic, but are also rhyolitic (i.e. having a granitic composition).

4.2.5 Ultra-high-pressure (UHP) metamorphism

During the 1960s and 1970s, the study of blueschist and eclogite-facies rocks established for the first time that some rocks of the continental crust have been buried to depths of around 50km and then exhumed. The scale of vertical movement was seen at the time as staggering. Imagine, then, the near-disbelief when metasediments exhumed from twice that depth were reported in 1984 by the French geologist, Christian Chopin, from the Dora Maira region of the Italian Western Alps. Here, rocks known to have experienced blueschist-facies conditions at relatively high temperatures were found to contain several indicators of extreme pressure, including the mineral **pyrope** (Mg-garnet) and, most notably the mineral **coesite**. Coesite was found as inclusions trapped within other grains (Fig. 4.19). Regional metamorphism that produces coesite is called **ultra-high-pressure (UHP) metamorphism**.

Coesite is a dense polymorph of SiO_2; it has the same composition as quartz. Experiments have established that its stability field lies on the high-pressure side of the line shown in Figure 4.20. Thus, for a plausible geotherm (temperature–depth

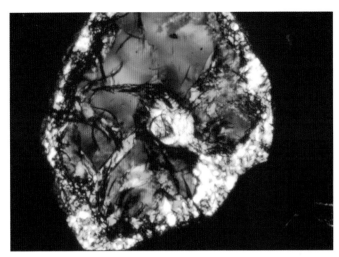

Figure 4.19 A thin-section photograph in XP showing a grain of coesite (SiO2) about 1mm in diameter enclosed by a large grain of garnet (black) in eclogite from an ultra-high-pressure metamorphic region. The tiny, bright yellow grain in the centre is pyroxene. The outer margin of the coesite, next to the garnet host, has inverted (changed) into fine-grained quartz, which appears in paler shades of grey than the coesite.
Source: Wikipedia commons https://commons.wikimedia.org/wiki/File:Coesiteimage.jpg Attributed to J. Smyth.

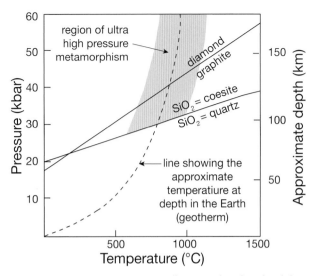

Figure 4.20 Pressure–temperature diagram showing, in pink, the field for ultra-high-pressure (UHP) metamorphism. It is in the stability field of coesite (a high-pressure form of SiO_2), and close to the estimated geotherm (curved dashed line).

curve) inside the Earth, shown as a dashed line in Figure 4.20, the pressure experienced by the Dora Maira rocks would have been at least 28–30kbar, corresponding to depths of more than 90 or 100km.

Ultra-high-pressure metamorphism has now been recognized in rocks from about twenty different parts of the world. Some of these places are geographically extensive, notably the so-called Western Gneiss Region of Norway, which is more than 30,000km² in area. This location, incidentally, is the source of the beautiful eclogite shown in Figure 2.54, and also the dunite in Figure 2.71.

In a few places with remnants of UHP metamorphism, such as the Kokchetav Massif in northern Kazakhstan, micro-diamonds, as well as coesite, have been reported, implying pressures in excess of 40kbar (equivalent to a depth of more than 120km), and temperatures of more than about 1000°C (Fig. 4.20). Microdiamonds are very difficult to spot because they are so small (only a few microns, which is much less than the thickness of a thin section) but, following their discovery in

the Kokchetav Massif, diligent searching elsewhere has found them in several other UHP regions, in all the continents. Moreover, other rare minerals have been recognized in UHP regions that point to even higher pressures – 60 to 90kbar – showing that rocks of the continental crust have been down to depths of over 200km before coming back to the surface.

All UHP metamorphic rocks have undergone very extensive retrograde metamorphism during their exhumation. Apart from inclusions of diamond and coesite, little survives of the primary high-pressure mineral assemblages. The coesite, too, would probably have changed to quartz if it had not been protected inside larger grains of other minerals. It survives, it seems, because it was physically armoured by strong minerals, with no room to expand. Also, it is likely to have been sealed against fluids that could have promoted recrystallization. The coesite inclusion illustrated in Figure 4.19 is fairly typical; it has partially altered to fine-grained quartz around its margins.

The only plausible mechanism for UHP metamorphism is subduction of continental crust on a plate that descends deep into the mantle (see Appendix 1, Fig. A1.5) before, presumably, upward buoyant forces overwhelm the down-drag and the UHP rocks become detached from their plate and rise rapidly.

Most examples of UHP metamorphism are found in orogenic belts of Phanerozoic age (less than 542 million years old). The continental crust today is 90km thick at most. Since some UHP rocks descended to more than 200km, then the depth of subduction of continental crust in the past was, in some cases, much greater than it is today.

4.3 Shock metamorphism

Shock metamorphism, which is caused by giant meteorite impacts, was introduced in a short paragraph at the end of chapter 1. This section elaborates on the process and describes its products.

Shock happens when rock is suddenly made to move faster than sound waves (i.e. seismic **P-waves**) can travel through it. P-waves travel at up to 6.5km/sec in most rocks, whereas meteorites arrive at speeds typically of 20km/sec. With shock, the atoms in mineral grains are instantly pressed extremely tightly together under intense, but short-lived, pressures that are ten times to one hundred times higher than those of normal regional metamorphism, i.e. 50 to 500kbar instead of 5kbar. (Shock pressures are often quoted in gigapascals (GPa); 1GPa is 10kbar). Closest to the impact, where the shock pressures are highest, the rock will instantly vaporize, causing the explosion that excavates the crater. Further out the rock will melt. Further out again it will suffer cataclasis (mechanical disaggregation). The crater is emptied out in minutes, but almost immediately the steep walls collapse inwards, sometimes rising as a central peak, and the wall-rocks become mixed in a chaotic way with incandescent ejected material as it falls back, partly filling the initial crater with a kind of breccia called **suevite** (pronounced **soo**-av-ite).

4.3.1 The discovery of shock metamorphism

Shock metamorphism is a relatively new branch of metamorphism, dating from the 1960s. Its emergence as a discipline is mirrored in the history of scientific investigations at the Barringer Meteorite Crater in Arizona (Fig. 1.13). This is one of the best-preserved craters on Earth, having formed only about 50,000 years ago in a desert region. During the first half of the twentieth century geologists argued over whether the crater had an explosive volcanic origin or a meteorite impact origin. One supporter of impact was Daniel Barringer, a mining engineer who linked the crater's origin with chunks of iron-nickel meteorite found in the surrounding desert. He

believed that a very large and valuable mass of metal, with a diameter approaching that of the crater, lay buried beneath the crater floor, so he obtained mineral rights and spent a fortune drilling down in search of it, but to no avail. Only later did it become clear that the crater was large because of an impact explosion, and that the meteorite had largely vaporized.

In 1960 the impact theory was strengthened when a piece of sandstone ejected from the crater and picked up in the desert was found to contain the minerals coesite and stishovite. These are two different dense polymorphs (crystalline varieties) of SiO_2 that require enormously high pressures to form (as seen for coesite in UHP metamorphism), consistent with the pressures produced, albeit briefly, by impact-induced shock waves. The evidence that finally convinced any remaining sceptics still favouring a volcanic origin came in 1963, when Eugene Shoemaker of the United States Geological Survey demonstrated that features produced in shock-damaged rocks from nuclear bomb testing sites could be matched in detail with features seen at Barringer Crater. Today the crater is marketed as a major tourist attraction. It is owned by Daniel Barringer's descendants, who generously donate a significant fraction of the profits to support research on impact cratering and meteorites.

By now almost 200 impact structures have been identified on Earth, including craters and deeply eroded remnants of craters. The two largest, Vredefort in South Africa and Sudbury in Canada, are each more than 250km in diameter. They are also the two oldest, having been formed, respectively, at just over 2000Myr ago and close to 1850Myr ago.

4.3.2 Products of giant impacts

The shock-damaged rocks at an impact site, and in the excavated ejecta, display one or more of the distinctive features known as shatter cones, pseudotachylite, dense polymorphs of silica and carbon, planar deformation features in quartz and other minerals, mineral grains that have been turned into glass (called diaplectic glass), and large-scale melt sheets.

Shatter cones are an unusual fracture pattern in the form of aligned sets of long, nested, striated cones (Fig. 4.21). The tips of the cones were once thought to be directed towards the impact, but that interpretation is now refuted because examples of shatter cones have been discovered that point in many different directions within a single outcrop.

Figure 4.21 Shatter cones in sandstone at Sudbury, Ontario. The notebook is about 10cm wide. *Photo courtesy of Gavin Kenny.*

Figure 4.22 Zone of breccia in impact-shocked granitic gneiss near Vredefort, South Africa. The dark material injected between the blocks is pseudotachylite. Angular blocks of gneiss appear to be in place. Rounded blocks may have been transported. Width of field is about 5 metres. *Photo courtesy of H. Jay Melosh.*

Pseudotachylite was introduced in chapter 2 (Fig. 2.68) as a product of dynamic metamorphism. At impact sites it is a dark, flinty-looking rock that appears to have been injected as a liquid into cracks that opened up in the host rock. The injections vary from broad dyke-like sheets down to millimetre-wide veins. In some large sheets, angular and rounded fragments of the host rock are embedded in the pseudotachylite, giving a breccia (Fig. 4.22). Here the pseudotachylite was produced *in situ* apparently by frictional melting as adjacent blocks of rock were jostled and shaken violently against each other.

The dense high-pressure polymorphs of SiO_2, coesite and stishovite, originally found at Barringer Crater have since been found at many impact sites, but identifying them is not straightforward and requires careful analysis using XRD. Also, shock-generated diamond has been found in carbon-bearing crater rocks. Coesite and diamond are not, of course, unique to shock metamorphism. As noted in the previous section, they are known in ultra-high pressure metamorphic rocks.

Planar deformation features in quartz are believed to be diagnostic evidence of shock metamorphism. They are parallel sets of dark lines visible under a microscope in individual grains or in a thin section (Fig. 4.23). Known simply as PDFs, they are found not just at impact sites, but also in tiny fragments of ejecta from distant craters. They are found, for

Figure 4.23 A shocked quartz grain, 0.32mm across, seen in XP through a polarizing microscope. It shows two strong sets of planar deformation features that partly overlap. This grain is from deposits within the Chicxulub crater, recovered from the drill hole Yucatán-6, located 50km from the centre of the crater. *Photo courtesy of Alan Hildebrand.*

example, in the worldwide 'boundary clay' that comprises fall-out from the 66Myr old Chicxulub crater in Mexico. This crater, at 180km in diameter, is the third largest on the planet, but remains hidden beneath younger sedimentary layers. Its discovery in 1990 can be attributed to serendipitous action by a geologist named Alan Hildebrand. It is implicated by many as the principal cause of the mass extinction event at the end of the Cretaceous Period that saw the demise of the dinosaurs.

Shock-melted minerals are another impact-related feature. In this case, the shock pressure and associated heating are sufficiently intense to cause total melting of individual mineral grains leaving, for example, pure silica glass or pure feldspar glass after the shock wave has passed. The glass is known as diaplectic (pronounced dye-a-**plek**-tic) glass. **Diaplectic glass** formed from plagioclase is known as maskelynite (pronounced **mass**-ke-lin-ite).

Finally, in very large impact craters such as those at Sudbury and Vredefort, the heating was so intense that a lake of melted rock, known as a melt sheet, was produced. At Sudbury it cooled slowly, and is now an enormous layer of coarse-grained igneous rock. Since the melting happens almost instantly, much of the liquid gets ejected as spray, along with pulverized and shock-damaged rock and mineral fragments, in the impact plume. The liquid coalesces into droplets that freeze to glass and fall back to the Earth's surface, possibly hundreds or even thousands of kilometres from the crater, as glass beads known as **tektites**. Small spherical tektites are called spherules. Thin layers of spherules, interleaved with normal sedimentary layers, have been reported from many places. Each of these so-called **spherule beds** is a testimony to some ancient far-off impact event. An example is shown in Figure 4.24, which is a thin section of a spherule bed of late Triassic age in south-west England, near Bristol. The spheres, which are up to 1mm across, are no longer glassy, having been altered to secondary minerals, including green chlorite. They are cemented by a form of K-feldspar, which has been dated by the ^{40}Ar–^{39}Ar method. The age is consistent with the spherules having been derived from the Manicouagan impact crater in eastern Canada, which is now deeply eroded. During the Triassic, the north Atlantic Ocean had yet to open up, so the impact site would have been much closer than it is today.

4.3.3 Extra-terrestrial shock metamorphism

While shock metamorphic effects are preserved in only a tiny fraction of the Earth's rocks, they are widespread on other bodies in the Solar System, where, in the absence of water and an atmosphere, and without plate tectonic recycling, shock

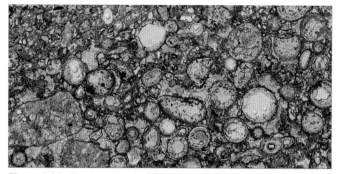

Figure 4.24 Thin section in PPL of the Triassic Meteorite Impact Deposit in the UK Virtual Microscope collection. The spherules, now altered to chlorite and other secondary minerals, are believed to have originated as glass droplets from the Manicouagan Impact Crater in Canada.

damage does not get healed. Giant impacts have affected the Moon, Mercury, Mars and most asteroids.

Meteorite craters are well known on the Moon. The lunar surface seen from Earth appears blotchy, with large rounded or irregular dark areas surrounded by light areas. These are known, respectively, as the lunar maria ('seas') and the lunar highlands. The maria are covered in smooth basalt lava flows with few craters, but the lunar highlands are very heavily cratered. Consistent with the cratered topography, samples collected from the highlands during the Apollo programme were found to consist almost entirely of breccia, produced by repeated large impacts into the original rock of the lunar surface. This rock is the igneous rock known as **anorthosite** (pronounced an-**or**-tha-zite) which is made largely of Ca-rich plagioclase, making it light in colour. A sample of anorthosite numbered 60025 from the Apollo 16 landing site is one of the oldest lunar rocks known. A thin section of it in XP in Figure 4.25 shows obvious cataclastic damage in a single large grain of plagioclase; the twin lamellae (stripes) have been off-set along miniature faults, and the grain as a whole has been fragmented, with each small piece shifted a little out of alignment with its neighbour. The overall effect is a little like a mosaic, so the plagioclase is said to have been mosaicized. In another sample, 78235 from the Apollo 17 site, shock has caused *localized* melting of plagioclase (Fig. 4.26). The melt (now maskelynite glass) has a swirling blue and brown pattern in PPL, and is isotropic in XP. Some unmelted but mosaicized (highly strained) plagioclase just beside the glass implies

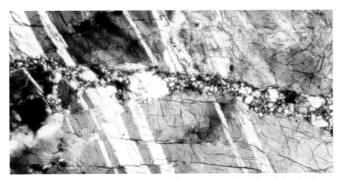

Figure 4.25 Thin section in XP of lunar anorthosite (plagioclase rock) 60025, showing part of a large mosaicized grain of plagioclase traversed by a band of finely crushed plagioclase.

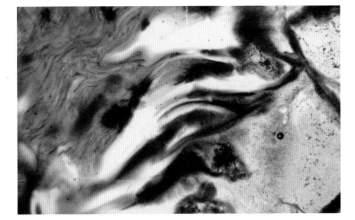

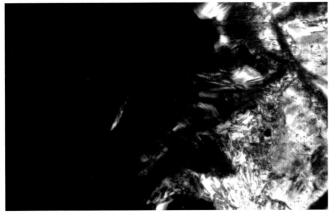

Figure 4.26 Shocked anorthosite in Apollo sample 78235 in PPL (top) and XP. For a description and comment see the text. Width of field is 2mm.

that the melting process, and hence the effect of the passing shock waves, was local and arbitrary. Almost all the cratering of the Moon's surface happened more than 4000Myr ago. The circular maria are actually giant impact basins, subsequently flooded with basalt lava.

Meteorites also show evidence of shock. They fall into a few dozen distinctive groups, and each group is thought to come from a single asteroid source. One group, known as the L chondrite group, includes many meteorites that have suffered intense shock. Figure 4.27 shows a thin section image of one of these meteorites in which the shock caused local melting. Fragments enclosed within the resulting dark glass contain clear olivine that has been partly transformed to a blue mineral called **ringwoodite**. Ringwoodite has the same composition as olivine; it is a dense high-pressure polymorph of $(Mg,Fe)_2SiO_4$. The transformation of olivine to ringwoodite is attributed to enormously high shock pressures caused when the L-chondrite asteroid collided catastrophically with another asteroid and broke up. The collision happened around 470Myr ago based on potassium–argon dating of shocked L-chondrites. Heat from the shock event drove off the argon gas that had been accumulating until that time in the meteorite's minerals,

Figure 4.27 Thin section in PPL of a pale-coloured olivine-rich fragment containing blue crystals of ringwoodite enclosed in dark, shock-melted glass in a meteorite called Taiban. Ringwoodite has the same composition as olivine. It is a high-density polymorph of $(Mg,Fe)_2SiO_4$. Width of field is 0.5mm. *Photo courtesy of Ed Scott.*

and the argon clock was re-set to zero. The collision launched millions of L-chondrite fragments into the asteroid belt, and soon afterwards the rate of arrival of L-chondrite meteorites on Earth jumped dramatically; a large number of 'fossil' L-chondrites have been found embedded in an Ordovician limestone in Sweden that was deposited soon after 470Myr ago. It is pure coincidence that this is also the time of peak metamorphism in the Grampian orogeny.

Ringwoodite, incidentally, is named in honour of A.E. (Ted) Ringwood, an eminent Australian geologist who was a pioneer in experimental petrology at high pressure. He led the way to the present understanding the nature of the **Moho** (see Appendix 1), the structure of the mantle, the origin of the Moon, and the production of basalt magma. He died in 1993 at the age of only 62. Those who knew Ted Ringwood speak of him with affection and with admiration for his prescient scientific insight.

5   Case studies in geothermobarometry

Exactly how hot do rocks become, and how deeply are they buried during metamorphism? Figuring out the so-called 'peak' temperature and pressure conditions is a major preoccupation among metamorphic geologists. The estimation of temperature is called **geothermometry**, and that of pressure, **geobarometry**. The rather cumbersome term **geothermobarometry** has been coined to cover both together.

This chapter discusses the application of geothermobarometric methods to rocks that were formed under unusually high-grade conditions in two separate places within the Grampian orogenic belt, one in Ireland and one in Scotland. The Irish example is from the SE margin of the belt, at a place called Slishwood. Here much of the Grampian belt is buried under a veneer of younger Carboniferous sedimentary rocks (Fig. 5.1). The Scottish example lies east of the Moine thrust on the NW margin of the Grampian belt near the coastal village of Glenelg (Fig. 5.15).

In both places the high-grade rocks existed before the Grampian orogeny, and were reworked (metamorphosed again) during the Grampian orogeny under lower-grade, amphibolite-facies conditions. Water gained access, causing widespread retrograde changes. Also, dynamic metamorphism occurred in discrete shear zones where the rocks were locally transformed to mylonite or blastomylonite. However, a good deal of the original high-grade 'dry' and 'uncrushed' rock survives, preserving pre-Grampian mineral assemblages and textures. It is these survivors that are now interrogated regarding the pressures and temperatures they experienced.

5.1 Granulite-facies rocks at Slishwood

The temperature and pressure at which the Slishwood rocks became metamorphosed have been estimated using four separate indicators. These are (1) the existence of kyanite, showing that the rocks formed within the P–T field where kyanite is stable, (2) an unusual texture in feldspar, known as perthitic texture, which provides evidence for high temperatures, (3) the combination of garnet, clinopyroxene and plagioclase in metabasites, which helps to constrain the estimate of pressure, and (4) the Fe/Mg ratios of garnet and coexisting clinopyroxene, which depend on temperature. The four approaches will be presented after first describing the geological setting of the Slishwood area.

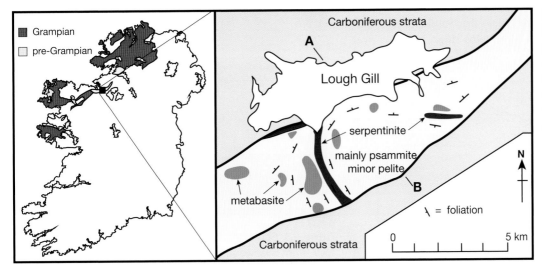

Figure 5.1 The location and a simple geological map of the Slishwood region and its surroundings. A vertical cross-section between the positions marked A and B is shown in Fig. 5.2.

5.1.1 Geological setting

Slishwood is the name of a townland (small area) at the heart of a picturesque region immortalized in the poetry of W.B. Yeats. It lies on the southern shore of Lough Gill, a few kilometres south-east of the city of Sligo on the Atlantic seaboard of north-west Ireland (Fig. 5.1). The rocks there stand out topographically as a narrow ridge of barren hills that runs in a NE–SW direction for more than 40km. The ridge is named the north-eastern Ox Mountains. The rocks are mostly psammitic gneisses and they protrude through a covering of younger (Carboniferous) sedimentary rocks with a faulted contact on the northern side (Fig. 5.2). Grampian reworking in the area around Slishwood is minimal.

The psammitic gneisses contain feldspar and garnet, and have a near-vertical foliation. The foliation is shared by the orientation of, and subtle colour banding within, a layer of serpentinite in the metasammites. The serpentinite layer has been preferentially eroded leaving a valley, called Slishwood Gap, cutting through the ridge (Fig. 5.1). Metre-sized lenses and larger masses of metabasite within the metapsammites are particularly abundant west of the serpentinite strip. They contain red garnet and green clinopyroxene, and might be mistaken for bodies of eclogite, but they also contain plagioclase, a mineral that is absent from true eclogite. Kyanite-bearing garnet-rich metapelite layers are also present locally. The rocks are almost bone dry here, perhaps because any water introduced during the period of Grampian reworking was soaked up by peridotite, converting it to the layer of serpentinite, leaving the other rocks dry. Similar rocks occur elsewhere along the NE Ox Mountains ridge, and also over a large separate region 40km to the north centred on Lough Derg in County Donegal, but all these rocks were affected by more extensive Grampian reworking.

5.1.2 Kyanite

Kyanite can be seen in Figure 5.3, which is the VM GeoLab garnet-kyanite gneiss, M07 from Slishwood. This rock is an augen gneiss, similar to the one collected by Charles Darwin from Rio de Janeiro (Fig. 2.69), but it contains kyanite and a good deal more garnet. Figure 5.3 shows its blastomylonitic texture (the texture of a mylonite that has partially recrystallized) with a large porphyroclast of kyanite in the centre. The kyanite can be examined online at rotation 1. It is slightly deformed with almost straight extinction, and has grey to cream interference colours. Quartz forms elongated grains that connect up to make a quartz 'ribbon' along the bottom of the image. The fine-grained streaky matrix is of recrystallized feldspar and biotite (brown).

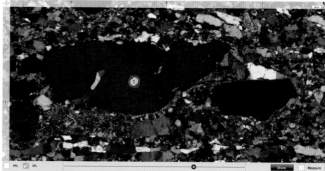

Figure 5.3 Screenshots of VM Geolab garnet-kyanite gneiss, M07, from Slishwood in PPL (top) and XP. It is described in the text. The field of view is 3mm wide.

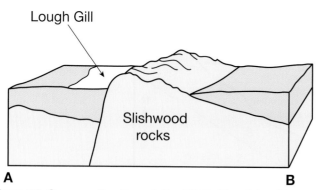

Figure 5.2 Cross-section through the NE Ox Mountains ridge at Slishwood from A to B in Figure 5.1.

The experimentally determined P–T field of kyanite (Fig. 2.32) is so large that one might regard it as of little value for narrowing down the P–T conditions for the Slishwood rocks, yet, when combined with the other P–T indicators about to be described, kyanite is of considerable value as will become clear.

5.1.3 Perthitic feldspar

Perthitic feldspar, or **perthite**, is an intricate mixture of tiny layers and lenses of albite (Na-rich plagioclase) and K-feldspar within a single feldspar grain. The two kinds of feldspar can be in any proportions. In some Slishwood rocks they exist in equal amounts (Fig. 5.4). Such perthite is called **mesoperthite**, and it implies unusually high metamorphic temperatures.

To understand its origin, one needs to know about **limited atomic substitution** in Na and K feldspars. This is explained in Appendix 2, and is explained again here.

Potassium feldspar ($KAlSi_3O_8$) usually contains some atoms of Na substituting for K, and albite ($NaAlSi_3O_8$) likewise usually has some K atoms substituting for atoms of Na. In both cases the amount of atomic substitution is limited because K^+ ions are much larger than Na^+ ions, and it is hard to squeeze a large K^+ ion into the small space normally occupied by Na^+,

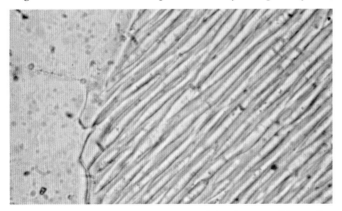

Figure 5.4 Mesoperthite in a thin section of psammitic gneiss from Slishwood at high magnification in PPL. The microscope stage has been lowered slightly, making the two kinds of feldspar distinguishable. The brighter strips are edge-on lamellae (layers) of albite; between them is K-feldspar. The uniformly clear mineral on the left is part of a very large grain of quartz that completely encloses the mesoperthite (see Fig. 5.6). Width of image is 0.2mm.

and a small Na^+ ion will rattle around uncomfortably in the large space normally occupied by K^+. Thus there is a gap in the compositional range between $NaAlSi_3O_8$ and $KAlSi_3O_8$ (see Fig. A2.3).

The width of the gap shrinks as the temperature rises, and eventually closes completely. The grey shaded area in Figure 5.5A shows the extent of atomic substitution at different temperatures based on experiments at a pressure of 5kbar. The limiting compositions define an arch-shaped curve that is called a **solvus**. The crest of the solvus is at 730°C, which is called the critical temperature.

To visualize the formation of perthite, one can imagine a protolith with equal amounts of Na and K feldspars being metamorphosed above the critical temperature. At this high temperature the rock will end up with just one kind of feldspar whose composition is in the middle of the range (position 1 in Fig. 5.5B). When the feldspar later cools it will find itself below the solvus curve, in the two-feldspar field, so it will no longer be stable. However, if it cools slowly enough, it will retain, or re-establish, a stable state by spontaneously splitting into a mixture of separate sodium-rich and potassium-rich feldspars, each with a composition lying on the solvus curve (position 2 in Fig. 5.5B). With even further cooling, the compositions of the two new feldspars will move further apart, staying on the solvus, until the temperature becomes too low (perhaps less than 500°C) for Na and K to move anymore (position 4 in Fig. 5.5B). This process is called **exsolution**, and the resulting intricate mixture of Na and K feldspars is mesoperthite.

The presence of mesoperthite at Slishwood clearly implies that the metamorphic temperature exceeded the critical temperature of the solvus. The critical temperature at 5kbar is 730°C but it increases by about 15°C for every kilobar increase in pressure. The position of the solvus at four different pressures is shown in Figure 5.5C. So the Slishwood mesoperthite implies that metamorphic conditions were in the grey area of Figure 5.5D, which is on the high-temperature side of a line marking the critical temperature of the solvus at different pressures.

By superimposing on this diagram the stability field of kyanite (see Fig. 5.10), it becomes clear that the kyanite field (shown in pale blue), and the grey area for P–T conditions above the solvus, overlap in a small wedge-shaped area (deep blue), which implies really rather hot and deep metamorphic conditions!

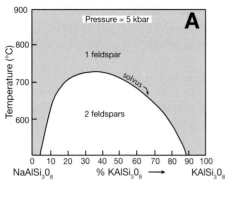

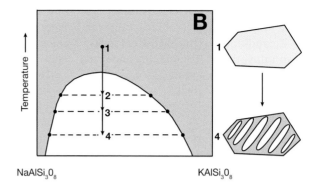

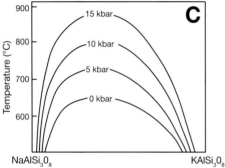

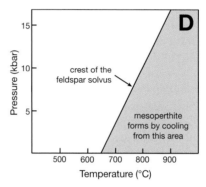

Figure 5.5 Diagrams relating to the formation of mesoperthite. **A**. The feldspar solvus at 5kbar. **B**. Cooling of homogeneous feldspar from above the solvus, leading to exsolution and formation of mesoperthite. **C**. Positions of the solvus at four different pressures. **D**. Critical temperature of the solvus as it changes with pressure. For an explanation see the text.

An incidental observation is that the mesoperthite occurs in small lenses that are completely enclosed in large ribbons of quartz (Fig. 5.6). The lenses are aligned parallel to the ribbons and parallel to the steep foliation seen in the field. Each lens must once have been a single homogeneous feldspar, because the mesoperthite inside the lens has a consistent orientation of parallel lamellae. Therefore, the directed stress responsible for the foliation must have been active when the temperature exceeded the crest of the solvus.

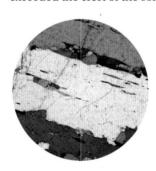

Figure 5.6 Screenshot of VM GeoLab Psammitic Gneiss, M06, from Slishwood in XP at rotation 1. It shows parts of three large grains of quartz (in different shades of grey) enclosing tiny parallel strips (lenses in 3D) of feldspar. The perthitic feldspar in Figure 5.4 is from a lens like one of these. Diameter is 2mm.

5.1.4 Garnet–clinopyroxene–plagioclase metabasites

An example of a garnet–clinopyroxene–plagioclase metabasite from Slishwood is shown in Figure 5.7. The combination of garnet, clinopyroxene and plagioclase is unusual for a metabasite. As noted above, it is like eclogite in having garnet and clinopyroxene, but eclogite contains no plagioclase. The presence of plagioclase places the Slishwood metabasite in a transitional position below the eclogite facies and in a high-pressure part of the pyroxene granulite facies, which some call the high-pressure granulite facies (Fig. 5.8).

The stability of the high-pressure granulite mineral assemblage has been investigated using the most common kind of basalt (called tholeiitic basalt) as an experimental starting material. This is very convenient because the Slishwood metabasites are tholeiitic, making the experimental results directly applicable to them. The stability field of high-pressure granulite was found to fall in an area bounded by a lower pressure of about 10kbar and an upper pressure of between about 12 and 14kbar. The stability field is shown later, in the

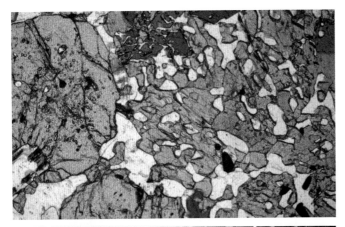

Figure 5.7 Screen shots in PPL (above) and XP of VM GeoLab garnet-clinopyroxene-plagioclase granulite, M23, from Slishwood. The image is from close to rotation 2. Two large, rounded grains of garnet are to the left. The pale green mineral in PPL with 'holes' is clinopyroxene; it is said to have a sieve texture. The 'holes' in the 'sieve' are plagioclase; the striped twin lamellae do not show here, but can be seen clearly elsewhere. The deep green mineral near the top is hornblende. Width of field is 2mm.

summary P-T diagram for Slishwood, Fig. 5.10. It is reassuring to note that the field is totally consistent with the conditions inferred from kyanite and mesoperthite.

5.1.5 Fe/Mg in garnet and in coexisting clinopyroxene

Clinopyroxene and garnet both contain iron and magnesium. However, the ratio Fe/Mg (strictly Fe^{2+}/Mg^{2+}) in garnet is

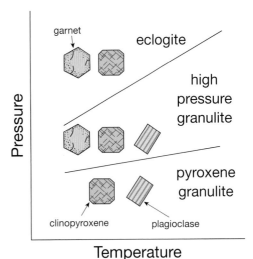

Figure 5.8 Cartoon showing how metabasite in the high-pressure granulite field, with garnet, clinopyroxene and plagioclase, is transitional between pyroxene granulite and eclogite.

different from the ratio in coexisting clinopyroxene. Iron goes preferentially into garnet, so Fe/Mg in the garnet will always be greater than Fe/Mg in clinopyroxene (Fig. 5.9, upper part). A parameter, with the symbol K_D, has been defined as the ratio of the two ratios, i.e:

$K_D = Fe^{2+}/Mg^{2+}$ in garnet divided by Fe^{2+}/Mg^{2+} in clinopyroxene

K_D is, therefore, a number greater than 1. What is important from the perspective of geothermometry is that K_D changes with temperature, and so provides the basis for what is called the **Fe–Mg exchange thermometer** for garnet and clinopyroxene. As the temperature increases and the atomic vibration becomes more energetic, Fe and Mg become less fussy about whether they reside in garnet or clinopyroxene, and so K_D gets smaller and closer to unity.

High-pressure experiments with tholeiitic basalt as a starting material have explored the variation of K_D with temperature (and also, to a lesser degree, with pressure). The results are summarized in the lower half of Figure 5.9 where a family of steep lines, each with constant K_D, show how K_D decreases in value as temperature rises.

K_D values have been determined for eight separate Slishwood metabasites. The chemical compositions of the garnet and clinopyroxene were measured using a technique known

94

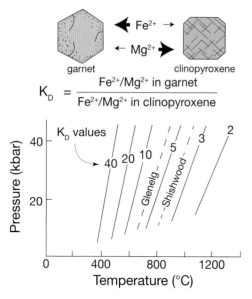

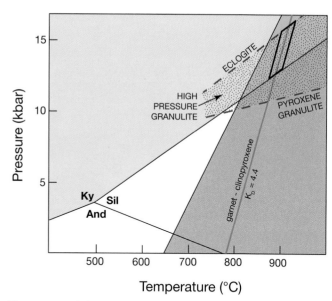

Figure 5.9 Fe–Mg exchange between garnet and clinopyroxene. It shows, at the top, how Fe preferentially goes into garnet; in the middle, the definition of K_D, and, at the bottom, the experimentally determined variation of K_D with pressure and temperature. For an explanation see the text.

Figure 5.10 Inferred pressure–temperature conditions for the rocks from Slishwood. The rocks formed where the blue area (kyanite field) overlaps the grey area (super-solvus K-Na feldspar field), in the stippled zone between the dashed red lines (high-pressure granulite field), and close to the line labelled $K_D = 4.4$. The best estimate is outlined as a bold-edged parallelogram, i.e. 12–15kbar and 850–900°C.

as **electron probe microanalysis**, which is described in Appendix 4.

K_D is reasonably constant in all eight samples, with an average value of 4.4 and a range from about 4.0 to 5.0. A line labelled 'Slishwood' with a K_D value of 4.4 is sketched on Figure 5.9. Again, it is consistent with the three pressure–temperature indicators discussed already.

Taking all four P–T indicators together (Fig. 5.10), it would appear that the Slishwood rocks were metamorphosed at a pressure of between 12 and 15kbar, and at a temperature of 850°C to 900°C.

The iron–magnesium exchange thermometer differs from the other three P–T indicators described above. The latter confine the P–T conditions to broad *areas*, and a good 'fix' becomes possible only if, through serendipity, the overlap zone of these areas is quite small. In contrast, the exchange of Fe^{2+} and Mg^{2+} between garnet and clinopyroxene constrains temperature and pressure conditions to a *line*. While at first the latter seems to be a better, more quantitative approach to estimating P–T conditions than to limit the conditions to

broad areas, it is not without its problems. The method has four obvious sources of error, which combine to give a potentially large error in the estimated temperature of perhaps ±50°C. Firstly, not all iron in pyroxene is Fe^{2+}; some of it is Fe^{3+}, which is not involved in the exchange with Mg^{2+} and has to be measured separately and removed from the total amount of iron (obtained using the electron microprobe) before calculating K_D. Accurate measurement of Fe^{3+} is difficult. Secondly, the positions of the K_D lines on Figure 5.9 are something of an estimate because they have been extrapolated from the results of experiments carried out at much higher temperatures. Thirdly, K_D is affected not only by temperature, but also by the presence of other elements in the garnet and clinopyroxene. Calcium in garnet is known to have a significant effect, and this has to be allowed for. Fourthly, the garnet and clinopyroxene grains may have compositions that are not uniform because Fe and Mg have continued to migrate between the two minerals while the rock was cooling but still very hot.

These problems aside, the exchange thermometer does appear to give plausible results for the Slishwood rocks and, as will be shown shortly, it also seems to give sensible results for the other suite of rocks to be considered here, those from Glenelg. Before looking at the Glenelg rocks, however, there is more of interest to report from Slishwood. This concerns the pressure–temperature *path* taken by the rocks before and after they reached their 'peak' P–T conditions.

5.1.6 Pressure and temperature trajectory

What can be said about the route taken by the Slishwood rocks from their original formation as sedimentary rocks at the surface, to the extreme metamorphic conditions at depth, just inferred, and then back to the surface where they are today? A limited answer comes from three clues found in the textures of the rocks.

The first clue may come as a surprise. Having just learned of the importance of kyanite in interpreting the Slishwood P–T conditions, it turns out that a few rocks contain sillimanite in addition to kyanite. The relationship between the two polymorphs of Al_2SiO_5 is significant. The sillimanite occurs locally in narrow quartz-feldspar-rich bands that cut diagonally across the main foliation; the bands appear to follow small faults or shear zones that moved after the main foliation had been established. Bundles of randomly radiating sillimanite prisms are enclosed within the quartz. The prisms in each bundle have been partially, or completely, replaced by one or two single grains of kyanite that faithfully replicate the shape made by the radiating sillimanite bundle (Fig. 5.11). The kyanite is said to pseudomorph the sillimanite. Each grain of kyanite must have grown from a single nucleus, and its replacement of sillimanite clearly took place in the complete absence of directed stress. Thus, at some rather late stage in their high-grade history, after the main foliation had been established, the rocks at Slishwood must have been in the sillimanite field, then moved from there into the kyanite field.

The second clue lies in the distinctive **sieve texture** of clinopyroxene with its inclusions of plagioclase and hornblende. An identical texture and mineral assemblage, admittedly with much smaller grains of plagioclase and hornblende, is well known elsewhere. It commonly occurs as a partial replacement of omphacite (Na–Al-rich clinopyroxene) in eclogite, where it is known as clinopyroxene–plagioclase **symplectite**. It occurs, for example, in the eclogite from Glenelg, where

Figure 5.11 Screen shots in PPL and XP of VM Geolab rock M22 from Slishwood showing a kyanite pseudomorph replacing a random bundle of sillimanite prisms in quartz. In XP the kyanite is in extinction (black) showing it is a single crystal. The blue grain, lower right, is a surviving end of a sillimanite prism. Most of the small, elongate grains with creamy yellow interference colours are also sillimanite.

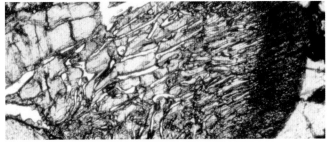

Figure 5.12 Thin section in PPL of a plagioclase-hornblende-clinopyroxene symplectite (intergrowth) replacing part of a large grain of omphacite in an eclogite from Glenelg. Note its similarity to the sieve texture of clinopyroxene at Slishwood, seen in Figure 5.7. The field of view is about 0.5mm.

it may form discrete rounded areas in the omphacite (Fig. 5.12). The symplectite is widely interpreted as evidence of

96

decompression, or possibly of heating, of eclogite. Either way, there can be little doubt that the metabasite at Slishwood had a former existence as eclogite, and the whole body of rocks must have moved during its subterranean odyssey *en masse* from the eclogite field to the high-pressure granulite field.

The third clue is the presence in some metapsammites of narrow borders of K-feldspar, locally accompanied by garnet, between kyanite and quartz. Figure 5.13 shows a K-feldspar border without garnet. The borders suggest that the kyanite and quartz reacted together, and that K-feldspar and garnet are products of the reaction. However, it is not possible to write a balanced chemical reaction involving just these four minerals. One interpretation is that another mineral, biotite, was also involved and has disappeared as a result of the reaction. A possible reaction would be:

$$K(Mg,Fe)_3Si_3AlO_{10}(OH)_2 + 2SiO_2 + Al_2SiO_5 = KAlSi_3O_8 + (Mg,Fe)_3Al_2Si_3O_{12} + H_2O$$

| biotite | quartz | kyanite | K-feldspar | garnet | water |

The breakdown of biotite in this way, like the breakdown of muscovite when it reacts with quartz, is usually taken to indicate increasing temperature. However, the reaction could instead have resulted from loss of pressure while the rocks were already at a very high temperature and being exhumed.

The three pieces of evidence for the direction of change in pressure and temperature are, at first sight, inconsistent. Taken at face value, the change from the sillimanite field to the kyanite field implies an increase in pressure and/or a drop in temperature, while the sieve-textured clinopyroxene and the inferred disappearance of biotite both imply the opposite sense of change. However, all three lines of evidence may be reconciled if the changes happened sequentially. Decompression (possibly with heating) could have happened first, to account for the sieve-textured clinopyroxene and the inferred loss of biotite. The decompression presumably brought the Slishwood rocks temporarily into the sillimanite field. Later cooling, while the rocks were still deeply buried, would account for the return to the kyanite field (Fig. 5.14).

This interpretation of the evidence for the evolving pressure and temperature of the Slishwood rocks is probably not unique, but it is a plausible scenario, and, reassuringly, it follows the 'standard' kind of P-T-t path, with heating during decompression, like the one shown in Figure 4.14.

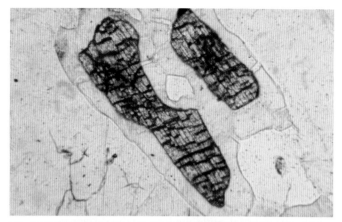

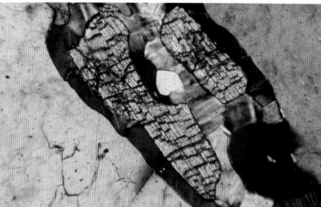

Figure 5.13 Thin section in PPL (above) and XP showing kyanite in psammitic gneiss from Slishwood. It is separated from quartz by a narrow border of K-feldspar. In PPL the microscope stage has been lowered slightly (as it was for showing mesoperthite in Fig. 5.4) causing the edge of the quartz to appear brighter and stand out. This use of the microscope allows the quartz and K-feldspar to be distinguished. It is known as the Becke test, and the bright edge is known as the Becke line. In XP the so-called tartan twinning of K-feldspar is visible. The pattern of two intersecting sets of cleavage can be seen in the kyanite, and is diagnostic, but only shows up in appropriately orientated grains, cut perpendicular to the cleavage. Width of field is about 1mm.

5.2 Eclogite-facies rocks at Glenelg

This section of the chapter presents a second example of the application of the garnet–clinopyroxene Fe–Mg exchange

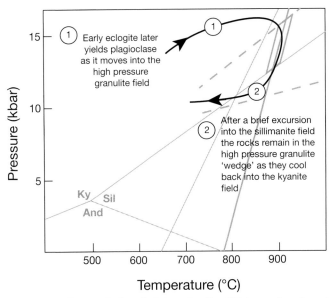

Figure 5.14 Suggested path taken by the Slishwood rocks at high pressure and temperature. The path is superimposed on Figure 5.10. For an explanation see the text.

thermometer, and it introduces two new methods, one based on the so-called calcite–dolomite solvus, and one based on limited atomic substitution of Na and Al in omphacite. The two new methods, like the garnet–clinopyroxene geothermometer, each constrain the estimation of P–T conditions to a line rather than an area.

5.2.1 Geological setting

The high-grade rocks in the Glenelg area are located in a remote strip of country within the Grampian belt immediately to the east of the Moine thrust. The strip extends for about 50km north to south centred on Glenelg village, and is shown in red in Figure 5.15.

During the Grampian orogeny, rocks in the Grampian belt were pushed upwards and westwards, sliding up the Moine thrust plane and onto a stable block of ancient continental crust to the west composed of so-called Lewisian gneiss. The Lewisian gneiss consists of orthogneiss (meta-igneous gneiss) and rarer metasediments that were metamorphosed more than 1700 million years ago.

The rocks in the Grampian belt are mostly much younger metasediments (psammitic gneisses and pelitic schists)

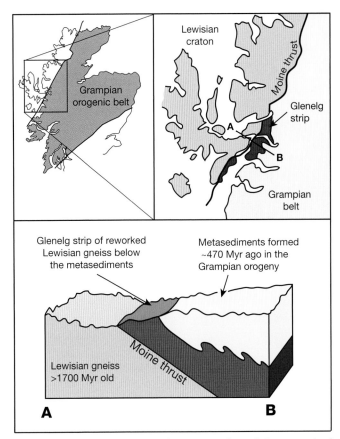

Figure 5.15 Location map and cross-section of the reworked Glenelg Lewisian rocks (in red) within the Grampian orogenic belt in NW Scotland. They are bounded by the Moine Thrust and are covered by younger metasediments (yellow). Original Lewisian rocks (not reworked) lie to the west (pink).

known as the Moine metasediments and coloured yellow in Figure 5.15, but the strip of high-grade rocks at Glenelg (coloured red) is thought to comprise Lewisian gneisses lying beneath the Moine metasediments that were reworked during the Grampian orogeny. The strip is referred to as the Glenelg Lewisian.

The Glenelg Lewisian is of interest because it includes many exposures of **eclogite**, which is otherwise very rare in the British Isles. Also present are outcrops of distinctive forsterite marble. Examples of both of these rocks were described in chapter 2 (Figs 2.38 and 2.55). Seen in outcrop, the marble

Figure 5.16 Part of a thick layer of eclogite in the Glenelg Lewisian (tinged red by garnet), which has been altered to amphibolite along shear zones (greyish green), presumably as a result of ingress of water during Grampian reworking.

is conspicuous with its protruding brown pebble-like grains of weathered forsterite (Fig. 2.37). The eclogite occurs as layers and lenses enclosed in grey quartz-feldspar-hornblende gneiss. It is rare to find eclogite that has not been partly altered to amphibolite (Fig. 5.16). This alteration is presumed to have occurred during the Grampian orogeny.

Another widespread feature is the development of zones of mylonite, which run the length of the Glenelg Lewisian strip parallel to the Moine thrust. The mylonite zones may have formed during the Grampian orogeny when dry Lewisian rocks, unable to recrystallize, succumbed to intense stress.

5.2.2 The calcite–dolomite solvus geothermometer

The calcite–dolomite geothermometer is applicable to marble containing both calcite and dolomite, such as the marble from Glenelg. It is based on the limited atomic substitution of Mg^{2+} for Ca^{2+} in calcite. The amount of substitution is limited because Mg^{2+} ions (radius 0.066nm) are considerably smaller than Ca^{2+} ions (0.1nm) so a Mg^{2+} ion would tend to 'rattle uncomfortably' in a site that would normally hold a large Ca^{2+} ion (see Appendix 2 for an account of ionic radii). The higher the temperature the more that Mg will enter calcite and substitute for Ca. The reciprocal substitution of Ca for Mg in dolomite also occurs, but to a far lesser extent. As with Na and K substitution in feldspars (Fig. 5.5),

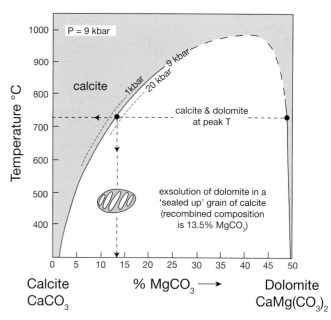

Figure 5.17 The calcite–dolomite solvus at 9kbar. For an explanation of its use in geothermometry see the text.

the increase in Mg substitution in calcite with temperature can be illustrated as a solvus curve (Fig. 5.17). This curve has been the target of several careful experimental studies, and its position on the calcite side of the solvus is known quite precisely at three different pressures. The curve for a pressure of 9kbar is highlighted in Figure 5.17.

Where marble contains calcite and dolomite in equilibrium, then the compositions of both minerals will fall exactly on the solvus curve. Assuming that the calcite retains its Mg during cooling, its composition will record the peak temperature, which can simply be read off the temperature axis of the solvus diagram.

A problem, however, is that Mg tends to escape from Mg-bearing calcite during cooling. This has happened in the Glenelg marble, where the composition of calcite is commonly about 6% $MgCO_3$, equivalent to a temperature of only about 550°C. This is presumed to be the temperature where movement of Mg ions becomes too sluggish for them to leave their calcite host.

Fortunately, in a few places the Mg did not get away. In these places tiny lamellae (plates) of dolomite grew inside

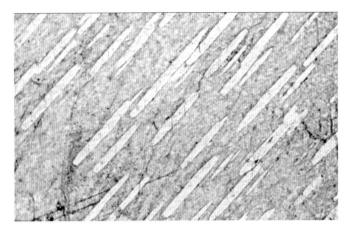

Figure 5.18 Thin section in PPL showing exsolved plates of dolomite, edge-on in a single grain of calcite (stained pink) trapped within a huge crystal of forsterite (not included in the image) from a sample of Glenelg marble. Width of field is about 1mm.

the calcite by the process of exsolution (in an equivalent way to exsolution in feldspar to make perthite). As the dolomite lamellae grew, they mopped up the Mg ions that were escaping from the adjacent cooling calcite. Thus the overall carbonate grain (residual calcite combined with the new dolomite lamellae) retained its full quota of Mg from when the calcite was at the peak temperature. An example is shown in Figure 5.18 – a calcite grain completely enclosed by a huge porphyroblast of forsterite (similar to the grain in Fig. 2.38). The composition of the original calcite can be obtained by measuring the compositions of the dolomite plates and of their calcite host, and measuring the volume percentage (which is the same as the percentage area in a thin section) of each. Such studies in marble from Glenelg consistently point to 13.5% of $MgCO_3$ (on a molecular basis) in the original calcite. This implies a metamorphic temperature of about 720°C at 9kbar. The temperature would be a little *lower* if the pressure were higher because the solvus decreases slightly with increased pressure (Fig. 5.17).

5.2.3 The clinopyroxene–albite–quartz geobarometer

The composition of clinopyroxene coexisting with albite and quartz serves as a geobarometer. In this case the composition changes with changing pressure (and also, to a small extent,

with temperature). The clinopyroxene in eclogite is a variety called omphacite, which has **coupled atomic substitution** between augite, $Ca(Mg,Fe)Si_2O_6$ and jadeite, $NaAlSi_2O_6$ (see Appendix 2). As Ca^{2+} is replaced by Na^+, so simultaneously (Mg^{2+}, Fe^{2+}) is replaced by Al^{3+}. The extent of this substitution depends on pressure because Al^{3+} ions are much smaller than Mg^{2+} and Fe^{2+}. Pressure compresses the large oxygen ions, making the octahedral spaces smaller, and more suitable for the smaller Al^{3+} ions. Na^+ has no problem swapping for Ca^{2+}; they are the same size.

If quartz is present with clinopyroxene, then there is a maximum amount of Al^{3+} and Na^+ (i.e. of jadeite) that can be accommodated, and it is this maximum value that changes with pressure. It is shown as a kind of 'pressure solvus' in Figure 5.19.

Eclogite is a metabasite which, by definition, never contains albite, so how can this geobarometer be applied? Fortunately, present among the rocks at Glenelg are some rare metadiorites that contain all three minerals. The omphacite in these rocks consistently has ~ 45% of the $NaAlSi_2O_6$ (jadeite)

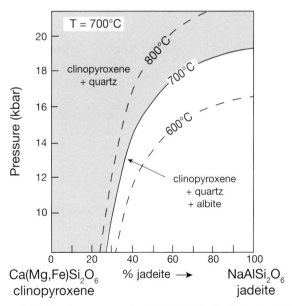

Figure 5.19 The 'pressure solvus' at 700°C for limited substitution (grey area) of jadeite in clinopyroxene, in the presence of quartz and albite. The approximate positions of the 'pressure solvus' at 600°C and at 800°C are also shown.

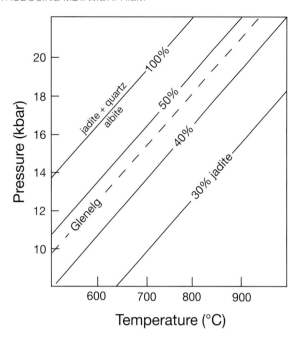

Figure 5.20 Experimentally determined pressure–temperature diagram showing the percentage of jadeite in clinopyroxene that coexists with quartz and albite. The Glenelg rocks formed under conditions, along the dashed line, for 45% jadeite.

end-member. This constrains the P–T conditions to a line on a pressure–temperature diagram (Fig. 5.20). The line has a relatively low slope, which crosses the steep line for the Mg-bearing calcite at a high angle and gives a good 'fix' on pressure and temperature (Fig. 5.21).

5.2.4 The garnet–clinopyroxene Fe–Mg exchange thermometer

The garnet–clinopyroxene Fe–Mg exchange thermometer that was used to estimate the temperature of metamorphism in the Slishwood rocks has been applied also in the Glenelg region. Ten mineral pairs were analysed. Their K_D values were found to be remarkably constant, with an average value of 6.4 ± 0.4. This is pleasing because there is an enormous range in Fe/Mg in the minerals analysed, and there is considerable variation in the size of the correction needed to deal with Fe^{3+} ions in the garnet and clinopyroxene. The K_D value of 6.4 ± 0.4 defines a line on a pressure–temperature diagram (Fig.

5.21, and also shown on Fig. 5.9) which passes right through the intersection of the lines corresponding to calcite with 13.5% $MgCO_3$ and omphacite with 45% jadeite, indicating that the Glenelg rocks were metamorphosed at about 700°C and 15kbar. The intersection of all three lines so precisely at one point must be, to some extent, fortuitous because each line has significant errors. A more realistic estimate of the P–T conditions is 700 ± 25°C and 15 ± 1kbar, and is shown by the orange ellipse in Figure 5.21.

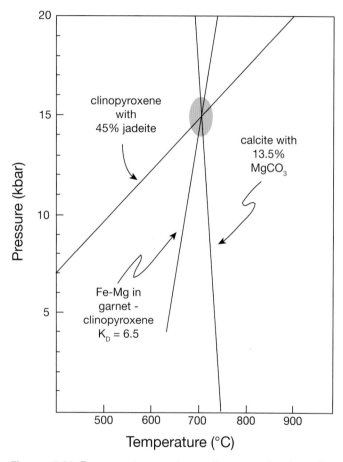

Figure 5.21 Pressure–temperature diagram showing the estimated conditions for the Glenelg high-grade rocks (orange ellipse). For an explanation see the text.

5.2.5 Dating the eclogite

An addendum to this chapter, unrelated to the issue of geothermobarometry, is the question of the age of the eclogite in the Glenelg Lewisian. For a long while, it was not known whether the eclogite was formed with the Lewisian gneiss to the west more than 1700Myr ago, whether it could be a product of Grampian metamorphism, or whether it was metamorphosed at some time in between.

Measuring the age of the eclogite is not easy. Grampian ages have been obtained widely from rocks of the Glenelg Lewisian, and these are taken to date Grampian reworking. Reworking tends to re-set isotopic clocks to zero, or to disturb them so much that they are rendered useless. However, an age of just over 1000 million years has been obtained using the so-called samarium–neodymium chronometer for garnet and omphacite. This isotopic clock is remarkably robust, and is not thought to be reset unless the temperature during reworking exceeds about 800°C. Neodymium (Nd) and samarium (Sm) are examples of Rare Earth Elements. They are widespread in rocks, but only in trace quantities. Nevertheless, technology today allows those tiny quantities to be measured precisely. The radioactive isotope samarium-147 (^{147}Sm) changes to the stable daughter isotope, neodymium-143 (^{143}Nd). As a result, the ratio of ^{143}Nd $/^{144}$Nd increases because ^{144}Nd is a stable reference isotope of Nd that does not change with time.

When the eclogite formed it is assumed that ^{143}Nd $/^{144}$Nd was the same in the garnet and clinopyroxene because they would have been in equilibrium. Garnet, however, always contains considerably more Sm relative to Nd than coexisting clinopyroxene. Therefore, over the passage of geological time, ^{143}Nd$/^{144}$Nd in garnet (high Sm/Nd) will have increased faster than ^{143}Nd$/^{144}$Nd in clinopyroxene (low Sm/Nd). The changes can be shown on a graphical plot called an **isochron diagram** (Fig. 5.22) where the vertical axis shows the present-day ratio of ^{143}Nd$/^{144}$Nd in garnet and in clinopyroxene, and the horizontal axis shows the ratio of ^{147}Sm$/^{144}$Nd in the same two minerals. When the eclogite formed, garnet and clinopyroxene would have been on a horizontal line (having identical

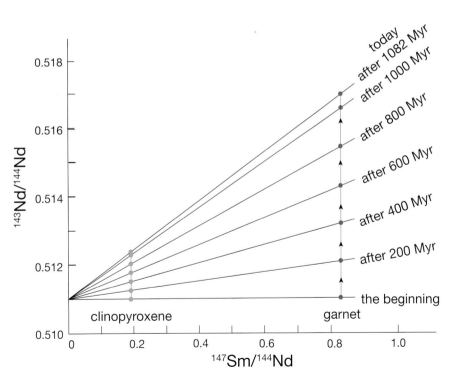

Figure 5.22 Samarium–neodymium isochron diagram for garnet and coexisting clinopyroxene in a sample of eclogite from Glenelg. The clinopyroxene (green dots) and garnet (red dots) began on a horizontal line with identical ^{143}Nd/^{144}Nd and moved upwards over time in proportion to their respective ratios of ^{147}Sm/^{144}Nd. They now sit on a sloping line (called an isochron) corresponding to an age of 1080 million years.

^{143}Nd/^{144}Nd). Through time, ^{143}Nd/^{144}Nd will have increased in both minerals in direct proportion to the ratio of ^{147}Sm/^{144}Nd in each mineral, so today the positions of the two minerals on the isochron diagram can be joined by a line whose slope is a measure of the eclogite's age. The position of the sloping line as it would have looked at intervals of 200Myr after the beginning shows how the present-day slope was achieved.

The age of just over 1000 million years was unexpected when it was first measured. No metamorphic rocks of that age had previously been reported from Scotland. It is much younger than the 1700 million years of the Lewisian gneiss to the west of the Moine thrust, and much older than the 470Myr Grampian event. It appears that the Glenelg Lewisian has a multi-stage history. Not only was it reworked during the Grampian orogeny, but the eclogite-facies metamorphism around 1000Myr ago was presumably also a reworking event that modified much older Lewisian gneiss. For some reason, the eclogite-facies metamorphism 1000 million years ago had no effect on the Lewisian gneiss to the west of the Moine thrust. Eclogite formation implies a major orogenic event, and such an event is known far to the east of Scotland in Scandinavia, and far to the west in Canada, where it is known as the Grenville orogeny. Why this orogenic event changed the Glenelg Lewisian, but failed to affect the rocks in the rest of the Lewisian, west of the Moine thrust, is not clear.

As a final comment relating to Sm–Nd dating of garnet-bearing rocks, and unrelated to the discussion of eclogite at Glenelg, it may be of interest to learn that the technique has now been refined to the point of measuring the *rate* at which a garnet porphyroblast has grown. This has been achieved by sampling an individual garnet porphyroblast at its core and at its margin using a micro-drill, and measuring the relevant Sm and Nd isotopes in the tiny volumes of garnet powder extracted. The measurements provide a date for core formation and a separate date for growth near the margin. Results from different metamorphic belts vary, and show that garnet growth took from a few hundred thousand years up to several million years. This kind of study gives a sense of the level of sophistication reached today in understanding, and quantifying, the thermal evolution of metamorphic rocks.

Appendix 1   The Earth's interior

Metamorphism is a process that takes place beneath the surface, often at considerable depth, so as a background to understanding how the process works, it is helpful to have a clear idea about the nature of the planet's interior. Important in this regard, and discussed in this short appendix, are the Earth's continental crust, its oceanic crust, the mantle beneath both, and the so-called tectonic plates that combine all three.

A1.1 The continental crust, the oceanic crust, and the mantle

The **continental crust** was discovered a little over 100 years ago by the Croatian seismologist Andrija Mohorovicčić. He carefully analysed many records of the time taken for sound waves (called seismic P-waves) to travel from an earthquake source to a recording station at various distances away, and found that the waves travel at speeds of up to 6.5km per second through rocks between the surface and a depth of roughly 35km, and travel at 8km per second through rocks at a depth greater than about 35km. An example of the kind of observation made by Mohorovičić is shown in Figure A1.1. The jump in speed from about 6.5 to 8km per second is now recognized beneath continents and shallow seas throughout the world. Rocks in the upper 35km or so, with P-wave velocities up to 6.5km per second, comprise the continental crust. The rocks beneath the continental crust comprise the **mantle**.

The boundary between the continental crust and the mantle is called the **Mohorovičič discontinuity** in honour of its discoverer. Most people just call it the **Moho** (pronounced mo-hoe). The depth to the Moho (i.e. the thickness of the continental crust) varies. The thickness of 35km, stated above, is only an average value. It is greater under mountain ranges, where it is locally as much as 90km, and can be 20km or less near the margins of continents.

The Moho has also been recognized globally beneath the deep ocean floor. The crust there is only about 7km thick on average, and is called the **oceanic crust** (Fig. A1.2).

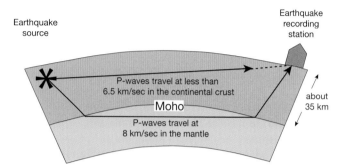

Figure A1.1 Cartoon, not to scale, of a vertical section through the outer part of the Earth in a continental region showing two paths taken by seismic P-waves travelling from an earthquake source to a recording station over a hundred kilometres away. At this distance, waves travelling via the mantle (green) arrive at the recording station before waves travelling directly, but more slowly, through the continental crust. The positions of the two arrow heads show the distances travelled by the P-waves in the same time.

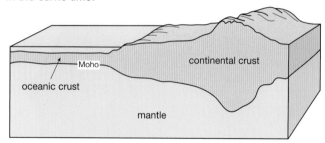

Figure A1.2 Cartoon to illustrate the thickness variation of the crust. The crust comprises the outer layer of the Earth, above the Moho, where seismic P-waves travel at 6.5km per second or less. It is typically about 35km under continents, much thicker under mountain ranges, but only about 7km under the deep oceans.

What is the crust made of? The continental crust is believed to consist of normal sedimentary, igneous and metamorphic rocks, based on the kinds of rock seen in very deep boreholes

104

and in places where the crust has been deeply eroded. The oceanic crust, in contrast, is believed to consist mainly of the dark-coloured igneous rocks called **basalt** (fine-grained), **dolerite** (medium-grained) and **gabbro** (coarse-grained) since these rocks are regularly recovered in deep-sea dredges and drill cores.

What is the mantle made of? It has not been possible to drill down through the continental crust to sample it. The deepest borehole (in the Kola Peninsula in north-western Russia) reached a depth of just over 12 kilometres, only about one-third of the way to the Moho. However, four lines of circumstantial evidence strongly suggest that the mantle is made of **peridotite**, a dense, green rock made largely from the mineral **olivine**. Olivine has the chemical formula $(Mg,Fe)_2SiO_4$ (this mineral formula is explained in Appendix 2).

Firstly, nodules (lumps) of peridotite are found in basalt (Fig. A1.3). Since peridotite is very rare in the crust, the nodules are believed to be detached fragments of the mantle through which the basaltic magma flowed rapidly on its ascent to the surface. Secondly, the speed of seismic P-waves in peridotite, measured experimentally, is about 8km per second, the same as the speed observed in the mantle. Thirdly, in a few parts of the world massive slabs of what appears to be oceanic crust and mantle, called **ophiolite** (pronounced **oh**-fia-lite) are exposed at the surface. The mantle part of an ophiolite

Figure A1.3 A broken lump of peridotite believed to have been brought up from the mantle entrained in flowing basaltic magma. A skin of black basalt is still attached to part of its surface. It consists mostly of large grains of the dense, green, glassy-looking mineral called olivine.

slab is made of peridotite. Fourthly, common meteorites, called **chondrites**, are made mostly of olivine. Chondrites are fragments of tiny planets (asteroids) from the asteroid belt, beyond the orbit of Mars. They have much the same chemical composition as the Sun (known from so-called absorption lines in the spectrum of sunlight), so they are thought to be made of the primordial (olivine-rich) starting material from which the Sun and the planets were made when the solar system was formed.

Why are the upper surfaces of oceanic crust and continental crust at different heights relative to sea level? The answer relates to density. Peridotite is about 3.3 times heavier than water (its density is 3.3 grams per cubic centimetre: g/cm^3), whereas most rocks in the crust are between 2.6 and 3 times heavier than water (average density 2.8 g/cm^3). Thus, the crust, whether continental or oceanic, is buoyant relative to the mantle, and 'floats' on top of it. The surface of most of the continental crust is close to sea level. Where the continental crust is thicker, it 'floats' with its surface above sea level, locally forming mountains whose base extends deep into the mantle. The oceanic crust is so thin that it 'floats' on the mantle with its surface well below sea level (Fig. A1.2).

A1.2 Plate tectonics

The continents are believed to have changed their positions over the course of geological time. Their movement is understood today in terms of the theory of **plate tectonics**. The theory emerged from the old hypothesis of continental drift, an idea proposed about a century ago by the German geophysicist and meteorologist Alfred Wegener. Wegener was fascinated by the way the opposing shorelines of the Atlantic Ocean appear to fit together almost perfectly, like pieces of a giant jig-saw puzzle. He imagined that the continents on either side were not only floating in the mantle, but were able to move through the mantle, like ice floes (sheets of floating ice). However, his ideas failed to gain acceptance, largely because geophysicists like Sir Harold Jeffreys in Cambridge pointed out that the mantle was simply far too strong to allow slabs of floating continental crust to drift through it in this way.

In the 1950s and 1960s it gradually became clear from studies of rock magnetism that Wegener and Jeffreys were both, in fact, right. The continents had, indeed, moved, but they had not moved *through* the mantle but had moved *with* the mantle. It was the mantle, or rather an outer layer of the mantle, that had moved.

This outer layer of mantle is now recognized as forming a planet-wide carapace of cool, strong rock about 100km thick, which is known as the **lithosphere** (meaning sphere of rock). Beneath the lithosphere, the mantle is hot and weak. Although it is solid it can easily be deformed and it behaves, over a long period of time, as a material that is mechanically weak and can flow very slowly, more like plasticine than rigid rock. This deeper part of the mantle has been named the **asthenosphere** (meaning 'sphere without strength').

The lithosphere is divided into separate pieces, like crazy paving, by deep cracks that extend through to the asthenosphere; it comprises a dozen or so 'paving slabs'. These slabs are not flat, of course, but curved to fit the Earth's spherical shape. They are called **tectonic plates**, or simply plates (*tectonic* comes from the Greek word *tekton-ikos*, meaning *to do with building*). There are eight large plates, four smaller ones, and several that are smaller still.

The plates are easily picked out on a world map of earthquake locations, because their edges are associated with earthquakes (Fig. A1.4). Earthquakes are caused because the plates are slowly sliding over the asthenosphere independently of each other, and their edges are repeatedly snagging, and being released in jolts.

The Earth's buoyant continental crust 'floats' in the mantle, just as Wegener imagined and as described above. It resembles several separate and extensive sheet-like rafts. However, in contrast to Wegener's ideas, the mantle in which the rafts 'float' is the strong lithosphere; the rafts are firmly embedded in it. As the plates move, so the rafts of continental crust simply ride along with them passively, like passengers on a travelator. Two of the eight large plates, named the Pacific Plate and the Nazca Plate, are covered entirely by oceanic crust. The other six plates carry both continental crust and oceanic crust, and each is named by the continent (or continents) within it (Fig. A1.4). The relative movement of plates means that they are converging in some places, and moving apart in others, as is

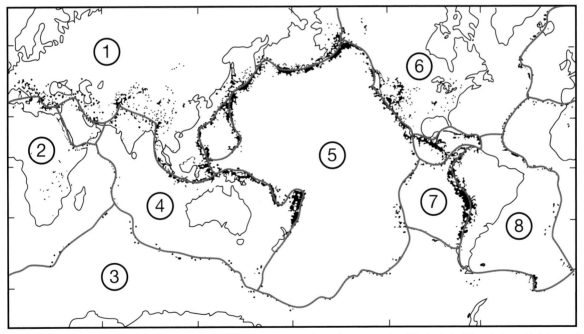

Figure A1.4 Map of the world with the locations of recorded earthquakes shown as tiny black dots. The dots, which have been 'joined up' with a red line, mark the edges of plates. Eight major plates are numbered, and named (1) the Eurasian, (2) the African, (3) the Antarctic, (4) the Australian/Indian, (5) the Pacific, (6) the North American, (7) the Nazca, and (8) the South American plates. Four minor plates (not numbered) are the Arabian, Philippines, Cocos, and Caribbean plates.

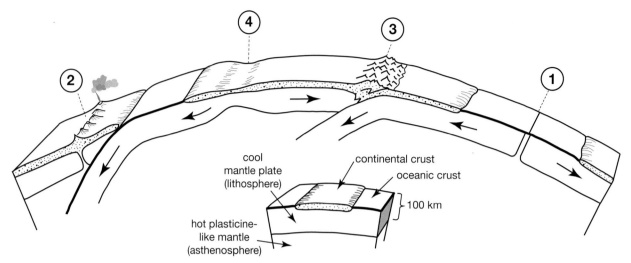

Figure A1.5 A hypothetical cross-section (not to scale) through the outermost 200 kilometres of the Earth. It includes parts of four lithospheric plates, about 100km thick. Oceanic crust is shown by a thick black line on the plate's upper edge; continental crust by a stippled outer capping. The plates are rigid and slide slowly over hot, weak mantle called the asthenosphere (see labelled inset drawing). Three kinds of boundary are numbered: (**1**) A mid-ocean ridge where plates are moving apart and new oceanic lithosphere is created. (**2**) A subduction zone where plates are converging and one of the plates, capped with oceanic crust, is subducted ('dives' at an angle into the mantle) beneath the opposing plate. (**3**) A subduction zone where both converging plates carry a cap of continental crust. (**4**) In a separate setting, not at a boundary, the lithosphere becomes stretched and thinned. More details of the processes at locations (1) to (4) are given in the text.

shown in Figure A1.5, which is a hypothetical vertical section through several plates and their mutual boundaries.

A1.2.1 What happens where plates move apart?

Where plates are moving apart, the hot asthenosphere creeps slowly upwards from a depth of over 100km to fill the widening gap between them (Fig. A1.5 position 1 and Fig. A1.6). As it rises, it partially melts to make a slush of basaltic liquid mixed with residual (unmelted) grains, mostly of olivine. Eventually the basaltic magma separates from the residual solids, rises towards the surface, then cools and hardens in the form of basalt pillows, dolerite dykes, and gabbro plutons, to make new oceanic crust. The mantle immediately below the oceanic crust consists of peridotite from which basaltic magma has been extracted. A line of volcanic activity known as a **spreading ridge** or **mid-ocean ridge** persists on the ocean floor between the diverging plates.

What causes the rising asthenosphere to melt? To answer this question one needs to know two things. Firstly, the

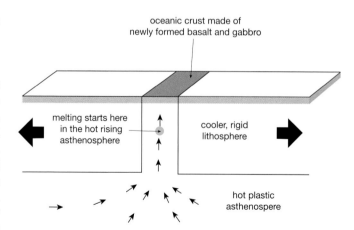

Figure A1.6 Cartoon showing the upwelling of hot asthenosphere to fill the widening gap between two plates that are moving apart. The asthenosphere starts to melt as it loses pressure, and the resulting basaltic magma separates out, rises further and cools to make new oceanic crust (shaded grey).

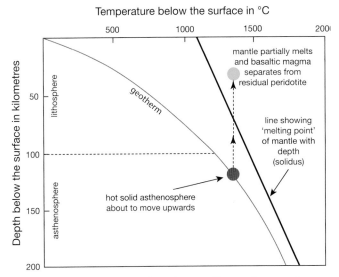

Figure A1.7 Diagram showing the geotherm (temperature v. depth curve) down to 200km and how basalt is produced by decompression when the mantle rises from 150km, crosses the solidus and starts to melt. For an explanation, see the text.

'melting point' of mantle peridotite has been measured experimentally. It increases enormously with pressure. At low pressure (shallow depths) it is about 1100°C, but at a depth of 200km, well within the asthenosphere, it is about 1800°C (Fig. A1.7). 'Melting point' is in quotation marks because mantle peridotite does not melt all in one go as it gets hotter, like ice for example. 'Melting point' is just the onset of melting, where the mantle first turns into a stiff slush. With increasing temperature, the stiff slush becomes runnier, and would become fully liquid if the temperature were to rise several hundred degrees more. The technical term for the line marking the 'melting point' is the **solidus**.

Secondly, one needs to know about the temperature in the mantle. The temperature going down a borehole on land rises typically by about 20°C for every kilometre of depth. However, it cannot continue to rise at this rate deep into the Earth; if it did, then 100km down the temperature would be around 2000°C, and the mantle would be totally molten. Yet the mantle is known to be solid at this depth because it can transmit seismic S-waves. **S-waves** are sideways shaking waves that are created alongside P-waves by earthquakes; they travel through solids but cannot pass through liquids.

Thus, it follows that the temperature in the mantle does not increase forever at a rate of 20°C per km, but follows a curved line that stays on the low-temperature side of the solidus in Figure A1.7. The curved line is called the **geotherm**.

At a depth of 120km, for example, the temperature on the geotherm is about 1400°C (marked by a red spot in Fig. A1.7). As the plates diverge, this asthenosphere remains hot as it rises to fill the gap, and will eventually cross the solidus and become partially molten slush. Basaltic liquid will then separate from the slush, as described earlier (green spot in Figs A1.6 and A1.7), and make its way further upwards to create oceanic crust. It is important to recognize that when basaltic magma is made in the mantle, the mantle is not heated up; it is already hot. It simply stays hot and starts to melt as it rises and loses pressure. The process is known as **decompression melting**.

Figure A1.7 shows something else of interest. It suggests why the lithosphere exists and why it changes to asthenosphere at about 100km depth. The geotherm near the surface is well below the solidus temperature, making the mantle here cool and strong, but it approaches close to the solidus at about 100km. Thus the mantle throughout the top 100km is strong, and comprises the lithosphere, but becomes less strong with depth. Below about 100km the geotherm remains close to the peridotite solidus, which makes the mantle here relatively weak (asthenosphere). The transition between the lithosphere and the asthenosphere is clearly not a sharp boundary like the Moho, but is gradual.

A1.2.2 What happens where plates converge?

Where plates converge, one of them takes a dive, sliding back into the hot mantle, as the opposing plate rides over it. The diving plate is said to undergo **subduction**. The region where it happens is called a **subduction zone**. There are two possible outcomes of convergence. They depend on whether the plate undergoing subduction is capped by oceanic crust or by continental crust.

If the diving plate is capped by oceanic crust then it slides back into the mantle quite easily, and a deep trench develops in the ocean floor where it starts to descend (Fig. A1.5, position 2). Large volcanoes erupt, typically about 100km or more beyond the trench. They form a so-called **volcanic arc**. The volcanic arc is in some cases built on top of oceanic crust, and the volcanoes are of basalt and also of the pale-grey rock andesite, which has a higher percentage of silica than basalt. In other cases, like the one shown in Figure A1.5, the volcanoes are built on continental crust, and they are made of pale-coloured rhyolite (from magma with a granitic composition) as well as from grey andesite and black basalt. This is the case, for example, among the volcanoes of the Andes, along the west coast of South America. Volcanic arcs suggest that the hot asthenosphere beneath them may be rising and partially melting, just as it is at mid-ocean ridges.

A very different picture emerges where the plate being subducted carries continental crust, as is depicted at position (3) in Figure A1.5. Enormous stresses build up in the down-going continental crust, because it is buoyant and resists descent. In its effort to stay afloat, as it were, it tends to pull away from its adjoining mantle, and to become intensely streaked out, buckled and bent as it piles up at the base of the overriding plate. If the overriding plate also carries continental crust, then this crust will combine with the crust below it, creating an enormously thickened continental crust that will result in a mountain range, as described in a simplified way in chapter 1 (section 1.2). A place where this process is going on today is beneath the Himalayan range, where the continental crust of India is being dragged inexorably down beneath that of Asia.

In some cases, continental crust is subducted beneath a plate with oceanic crust (not shown in Fig A1.5). When this happens the buoyancy of the continental crust resists subduction, which eventually stops, but not before the leading edge of the oceanic plate has ridden up over the down-going continental margin, where it comes to rest as a slab of ophiolite (as mentioned in section A1.1).

A1.2.3 Subsidence within plates

Subsidence of the Earth's surface takes place where the continental lithosphere becomes stretched and becomes thinner, but without actually being ruptured. When the lithosphere gets thinner, the buoyant upper layer of continental crust also gets thinner, so no longer 'floats' as high in the mantle (position 4 in Fig. A1.5). The surface of the continental crust sinks below sea level, creating a depression over a wide area known as an **extensional sedimentary basin**. Great thicknesses of sedimentary rock will accumulate in such basins. The North Sea is an example of a flooded extensional basin.

Appendix 2   The chemical formulae of minerals

Minerals are made from chemical elements. Over 90 different elements exist in nature, from the lightest element, hydrogen, to the heaviest, uranium. As is stated in chapter 1, only ten of these elements are needed to make the two-dozen or so minerals that comprise the vast majority of metamorphic rocks. These ten are hydrogen (H), carbon (C), oxygen (O), sodium (Na), magnesium (Mg), aluminium (Al), silicon (Si), potassium (K), calcium (Ca), and iron (Fe). Eight of these elements account for around 99% by weight of the **continental crust** (Fig. A2.1). The other two, hydrogen and carbon, are included in the tiny sector labelled 'others' in Figure A2.1. Two elements, oxygen and silicon, are especially abundant.

A2.1 How are chemical formulae of minerals written?

The atoms if elements combine together in simple proportions to make minerals. The chemical formula of a mineral is a kind of shorthand notation for writing down these proportions. For example, the formula for **quartz** is SiO_2. The subscript '2' means that there are two atoms of oxygen for every one atom of silicon. The mineral **kyanite** has the formula Al_2SiO_5, which shows that it contains aluminium, silicon and oxygen atoms in the proportions 2:1:5. The mineral **dolomite** has the formula $CaMg(CO_3)_2$. The brackets here, followed by the subscript '2', mean that there are two lots of '(CO_3)' for one atom each of calcium and magnesium. Thus (CO_3) must be multiplied by 2 to get the overall proportions of the elements in dolomite: calcium, magnesium, carbon and oxygen are in the proportions 1:1:2:6.

What fixes the numbers? Why is kyanite's formula written Al_2SiO_5, and not Al_3SiO_5? The reason is that elements occur as electrically charged atoms called **ions**, and not as neutral atoms. Ions have to combine in strictly defined ratios that are dictated by the electrical charge. Neutral atoms of any particular element, such as silicon, have a central nucleus with a fixed number of particles called protons (14 for silicon), each proton having one positive charge. The nucleus is surrounded by an equal number of minuscule particles called electrons

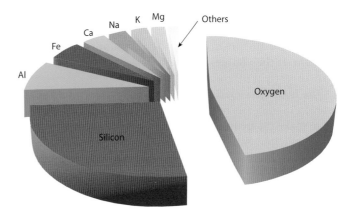

Figure A2.1 Proportions by weight of the eight most abundant elements in the continental crust. Two others, hydrogen and carbon, are included in 'others', and all ten are needed to make the common minerals in metamorphic rocks.

'orbiting' the nucleus, each with one negative charge. The neutral atoms become ions by spontaneously losing or gaining one or more electrons. Loss of electrons will leave an ion with more protons than electrons, so it will be positively charged. Conversely, gain of electrons will convert a neutral atom into a negatively charged ion. The charge is simply the number of electrons lost or gained; it is written as a superscript numeral with a plus or minus sign following the element symbol. The ten ions that are most important in minerals, in order of increasing mass, are:

H^+, C^{4+}, O^{2-}, Na^+, Mg^{2+}, Al^{3+}, Si^{4+}, K^+, Ca^{2+} and Fe^{2+}.

Only one of these elements, oxygen, occurs in the form of negatively charged ions. The other nine form positively charged ions. Since opposite charges attract, the positive ions are attracted to, and become tightly bound to, the negative oxygen ions by so-called **ionic chemical bonds** to make minerals.

In writing the chemical formula of a mineral, the sum of charges on the positive ions must be exactly equal to the sum of negative charges on the oxygen ions. Continuing with the example of kyanite, its formula, Al_2SiO_5, has two Al^{3+} ions and one Si^{4+} ion giving a total of 10 positive charges, and it has five O^{2-} ions that give 10 negative charges. The positive and negative charges are therefore balanced and overall the formula is neutral. To give another example, in the mineral calcite, whose formula is $CaCO_3$, Ca^{2+} and C^{4+} together provide 6 positive charges, while three O^{2-} ions give 6 negative charges. This simple process of balancing the negative and positive charges works for every mineral formula. If the charges do not balance, then the formula is incorrectly written.

The hydrogen ion, H^+, is not normally treated as a separate ion in minerals because it almost always teams up with O^{2-} to make the **hydroxyl ion** $(OH)^-$. So $(OH)^-$ is usually taken as a single ion with just one negative charge rather than as separate O^{2-} and H^+ ions. Another point to note is that iron atoms in nature can occur as Fe^{3+} ions as well as Fe^{2+} ions, by losing three electrons instead of two.

The number of charges on an ion can be worked out from its position the **Periodic Table**, which lists all the elements in the order of the number of protons that each element has in the atomic nucleus. The first element, hydrogen, has one proton per atom, the second, helium, has two protons per atom, lithium has three, beryllium has four, and so on, increasing by one proton per element. The list is arranged in a table with rows and columns. The number of protons increase going along the rows. Elements with similar properties lie in the same column. Elements in the first column have ions with one positive charge, those in the second column have ions with two positive charges, and so on. The Periodic Table for the first 20 elements, shown as ions, is set out in Figure A2.2.

To help memorize the Periodic Table, the element symbols can be strung together and read aloud as long 'words'. The first two 'words' sound like 'her-helly-beb-cernoff-knee' and 'nam-gall-see-puss-clar'.

A2.2 Minerals whose composition can vary

In many minerals the composition is not fixed, as it is in quartz, kyanite and calcite, but it can vary within limits. For example, the mineral **olivine** can have any intermediate composition between pure magnesium olivine, Mg_2SiO_4, and pure iron olivine, Fe_2SiO_4 (see Fig. A2.3-A). The formula of olivine is written $(Mg,Fe)_2SiO_4$ to indicate this possible variation in Fe and Mg. Iron and magnesium are said to substitute for each other, so the chemical variation is called **atomic substitution**. Olivine is described as a **solid solution series** between two end-members. The **end-members** have their own names. Mg_2SiO_4 is called **forsterite** and Fe_2SiO_4 is called **fayalite**.

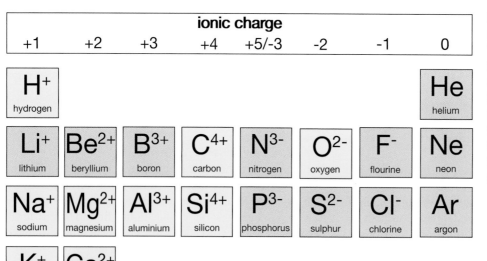

ionic charge							
+1	+2	+3	+4	+5/-3	-2	-1	0

H⁺ hydrogen — **He** helium

Li⁺ lithium | **Be²⁺** beryllium | **B³⁺** boron | **C⁴⁺** carbon | **N³⁻** nitrogen | **O²⁻** oxygen | **F⁻** flourine | **Ne** neon

Na⁺ sodium | **Mg²⁺** magnesium | **Al³⁺** aluminium | **Si⁴⁺** silicon | **P³⁻** phosphorus | **S²⁻** sulphur | **Cl⁻** chlorine | **Ar** argon

K⁺ potassium | **Ca²⁺** calcium

Figure A2.2 Periodic table showing the first 20 elements. The nine elements highlighted in yellow, along with element number 26, iron (Fe), are the ten important elements in minerals. The charge on each ion is determined by the column in which the element sits. Of the highlighted elements, only oxygen occurs as negative ions.

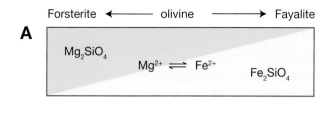

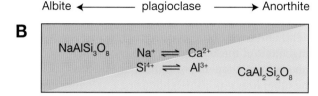

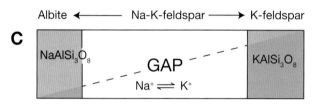

Ion	Radius in nanometres	Coordination number	Coordination polyhedron	Packing geometry
C^{4+}	0.015	[3]		
Si^{4+}	0.04	[4]		
Al^{3+}	0.05	[4] [6]		
Mg^{2+}	0.066	[6]		
Fe^{2+}	0.074	[6]		
Ca^{2+}	0.10	[8]		
Na^+	0.10	[8]		
K^+	0.14	[12]		

Figure A2.4 Radii of the main positively charged ions in common minerals and the way each is surrounded by a polyhedron of large, negatively charged O^{2-} ions (and also $(OH)^-$ ions). The number of oxygen ions in a polyhedron is called the coordination number of the positive ion at its centre. The smaller the radius of a positive ion, the smaller is its coordination number, and the larger is its charge.

Figure A2.3 Three styles of atomic substitution. (**A**) Simple substitution of Fe^{2+} for Mg^{2+} in olivine. (**B**) Coupled substitution of Ca^{2+} for Na^+ and Al^{3+} for Si^{4+} in plagioclase. (**C**) Limited substitution of K^+ for Na^+ and *vice versa* in Na- and K-feldspars.

Atomic substitution between Fe^{2+} and Mg^{2+} happens because the two ions are the same size. Ions behave like small rigid spheres with a definite radius. A list showing the radii of common ions is presented in Figure A2.4, where Fe^{2+} and Mg^{2+} are both seen to have a radius of about 0.07nm. One nm (nanometre) is one-millionth of a millimetre, so atoms are pretty small!

Another example of atomic substitution occurs in the mineral **plagioclase feldspar** (usually just called plagioclase), which is the most abundant mineral in the continental crust. Plagioclase is a solid solution series between the end-member called **albite**, $NaAlSi_3O_8$, and the end-member called **anorthite**, $CaAl_2Si_2O_8$, (Fig. A2.3-B). The ions Na^+ and Ca^{2+} each have a radius of about 0.10nm (see Fig. A2.4), but they have different charges. Nevertheless, they do substitute freely for each other because a second atomic substitution, of Al^{3+} for

Si^{4+}, takes place simultaneously, keeping the overall charge neutral. Al^{3+} and Si^{4+} ions have slightly different radii (see Fig. A2.4) but they are close enough for them to substitute for each other. This style of atomic substitution, where two separate substitutions take place in tandem to keep the charge balanced, is called **coupled atomic substitution**.

A third kind of substitution is called **limited atomic substitution**. It occurs, for example, in **potassium feldspar** (K-feldspar), $KAlSi_3O_8$. Here, Na^+ can substitute for K^+, but only to a limited extent because the two ions are of different sizes. The radius of Na^+ (about 0.10nm) is much less than the 0.14nm for K^+ (see Fig. A2.4). In the same way, there is limited substitution of K^+ for Na^+ in albite, $NaAlSi_3O_8$. Both these examples are shown together in Figure A2.4-C, where the range of compositions extends a short way out from each end, with a gap in the middle.

The width of the gap shown in Figure A2.4-C is not fixed, but depends on the temperature. The gap gets smaller (the limit of atomic substitution increases) as the temperature rises, and eventually the gap closes altogether (at about

700°C). The reason for this is because the whole crystal structure vibrates more vigorously at higher temperatures, and each kind of feldspar becomes less 'fussy' about accepting ions of the wrong size.

Limited atomic substitution is widespread among minerals, and since it depends on temperature, it has been used in estimating the temperature at which a rock is metamorphosed, as shown in chapter 5.

A2.3 How are atoms (ions) stacked together?

In all minerals the atoms are packed closely together in a pattern called a **crystal structure**, which is a pattern that is repeated regularly in three dimensions with negative and positive ions next to each other. The negative oxygen ions (and also the $(OH)^-$ ions) are larger than most others, having a radius of 0.14nm. They cluster together, as shown in Figure A2.4, to make 'cages' or **polyhedra** out of three, four, six, eight, or twelve atoms. A positively charged ion fits snugly into the centre of a polyhedron (Fig. A2.4). The number of oxygen atoms in a polyhedron is called the **coordination number** of the central positive ion, and it depends on the central ion's size.

Carbon, the smallest, has a coordination number of 3; it sits between three oxygens giving the **carbonate ion**, $(CO_3)^{2-}$. Silicon, Si^{4+} (radius 0.04nm) has a coordination number of 4; it fits nicely inside a **tetrahedron** (four touching oxygen atoms) so all silicate minerals contain $(SiO_4)^{4-}$ tetrahedra (the traditional plural of tetrahedron). Mg^{2+} and Fe^{2+} ions (radius about 0.07nm) need more space. They have a coordination number of 6, so are surrounded by six oxygens, making an octahedron (a polyhedron with six corners and eight triangular faces; Fig A2.4). Na^+ and Ca^{2+}, the two ions with a radius of about 0.1nm that substitute for each other in plagioclase, have a coordination number of 8. They fit between eight touching oxygens. K^+ is the same size as an oxygen ion, so it is surrounded by 12 oxygens in a large polyhedron.

Aluminium, Al^{3+} (radius of 0.05nm) is something of a misfit. It is bigger than Si^{4+} but smaller than Mg^{2+}. In many minerals, including feldspar, Al^{3+} ions go into a tetrahedron, while in some minerals, like kyanite, Al^{3+} is in an octahedron. Aluminium's dual behaviour gives rise to a style of coupled atomic substitution where two Al^{3+} ions simultaneously replace one Si^{4+} ion in a **tetrahedral site**, and one Mg^{2+} ion in an **octahedral site**. This kind of substitution is always limited. Octahedral aluminium is favoured by high pressure, because high pressure compresses the oxygen atoms and reduces the space at the centre of an octahedron, making it better suited for the small Al^{3+} ions.

A2.4 Classification and properties of silicates

Most of the abundant minerals in rocks are composed largely of the elements silicon and oxygen. They are known as **silicate minerals**, or simply as silicates, and they are divided into five main groups. All silicates have silicon–oxygen tetrahedra, and each of the five groups has its own distinctive arrangement of these tetrahedra. In the first group the SiO_4 tetrahedra are independent units, not touching each other. In the other four groups neighbouring tetrahedra are joined at their corners by sharing oxygen atoms. They are joined to make long single chains, or pairs of parallel cross-linked chains (double chains), or sheets, or three-dimensional frameworks. Single chains, double chains and sheets are illustrated in Figure A2.5.

A2.4.1 Silicates with independent tetrahedra

Seven of the common silicates have crystal structures with independent tetrahedra. They are olivine, garnet, kyanite,

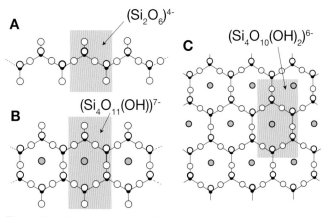

Figure A2.5 Arrangements of SiO_4 tetrahedra in (**A**) single chain silicates, (**B**) double chain silicates and (**C**) sheet silicates. Small black spheres are silicon atoms. Each is linked to four large white spheres, which are oxygen atoms. Green circles in (**B**) and (**C**) are hydroxyl $(OH)^-$ ions. In each group, the repeat unit is highlighted by a pink rectangle in which the numbers of atoms can be added up to give its formula and negative charge.

andalusite, sillimanite, staurolite and epidote. Each tetrahedron has four negative charges, written $(SiO_4)^{4-}$, so part of the mineral formula will be 'SiO_4' or a multiple of 'SiO_4'. The negative charges are balanced by various positive ions.

Olivine was described above: Mg^{2+} and Fe^{2+} ions provide the four positive charges and give the formula $(Mg,Fe)_2SiO_4$.

Garnet has several end-members with limited atomic substitution between them. The most common is iron garnet, called **almandine**, which has 2 lots of Al^{3+} with 3 lots of Fe^{2+} giving 12 positive charges and the formula $Fe_3Al_2(SiO_4)_3$ (also written $Fe_3Al_2Si_3O_{12}$). Rarer is calcium garnet, called **grossular**, whose formula is $Ca_3Al_2(SiO_4)_3$. A very rare magnesium end-member, $Mg_3Al_2(SiO_4)_3$, has the name **pyrope** (pronounced **pie**-rope) and only forms at extremely high pressures. Iron garnet usually has limited atomic substitution of Ca and Mg, and also of the less common element manganese (Mn).

Kyanite, andalusite and sillimanite all have the same formula, Al_2SiO_5, which can be written, alternatively, as $Al_2O(-SiO_4)$ to show the presence of the SiO_4 tetrahedron. In cases like this, where two or more minerals have the same chemical formula, the minerals are called **polymorphs** of that formula. Kyanite, **andalusite** and **sillimanite** are all polymorphs of Al_2SiO_5.

Staurolite's formula is quite complex, but it approximates to two lots of Al_2SiO_5 and one of $FeO(OH)$.

Epidote is a mineral with the formula $Ca_2(Al,Fe^{3+})Al_2(-SiO_4)_3(OH)$. Here iron exists as the Fe^{3+} ion rather than the Fe^{2+} ion, and it can substitute for up to one-third of all the Al^{3+} ions. In epidote some of the SiO_4 tetrahedra are not, in fact, independent, but linked in pairs.

In all the above minerals the atoms are efficiently packed, giving a high density (and high relief when viewed under the microscope – see Appendix 3). Most have rather poor **cleavage**. The efficient packing of atoms is exemplified by olivine. Here, layers of oxygen ions form a so-called **close-packed structure** (Fig. A2.6). The small spaces between adjacent layers include tetrahedral sites where 4-fold coordinated Si^{4+} ions fit, and octahedral sites where 6-fold coordinated Mg^{2+} and Fe^{2+} ions fit.

A2.4.2 Single chain silicates
Single chain silicates are also called **pyroxenes**. The chains have a repeat formula of $(Si_2O_6)^{4-}$ (see Fig. A2.5). They are

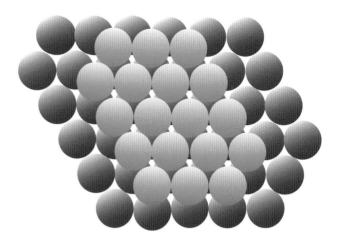

Figure A2.6 Two close-packed layers of touching spheres representing oxygen ions in olivine. Colours are used only to help distinguish the layers. A tetrahedral site is present below each orange sphere and also between three orange spheres and a purple sphere below. Si^{4+} ions occupy such sites. Octahedral sites (for Mg^{2+} and Fe^{2+} ions) exist midway between the two layers where the white background shows through.

parallel to each other, and off-set relative to each other in two different ways. In the first way (called **orthopyroxene**) the chains are positioned so that only Fe^{2+} and Mg^{2+} ions in 6-fold (octahedral sites) can easily be accommodated between them. The formula of orthopyroxene is therefore $(Fe,Mg)_2Si_2O_6$ (or $(Fe,Mg)SiO_3$ if you prefer). The Mg end-member, $MgSiO_3$, is named **enstatite**. It can occur with olivine in peridotite.

The alternative off-set positioning of the chains gives **clinopyroxene**. Here, equal numbers of 6-fold and 8-fold sites exist between the chains. In the simplest case Mg^{2+} ions go into the 6-fold sites and Ca^{2+} ions go into the 8-fold sites, giving the formula $CaMgSi_2O_6$. This is an end-member called **diopside,** which can occur in marble. The most common clinopyroxene is the mineral **augite,** which has a wide variety of limited atomic substitution, including Fe^{2+} for Mg^{2+}, Na^+ for Ca^{2+}, Mg^{2+} for Ca^{2+}, Al^{3+} for Mg^{2+}, and Al^{3+} for Si^{4+}. A pure sodium variety, $NaAlSi_2O_6$, is named **jadeite**, and is stable only at very high pressure because the aluminium is in the octahedral site. Clinopyroxene, with a composition between that of jadeite and augite, is called **omphacite**. It occurs in the metamorphic rock, **eclogite.**

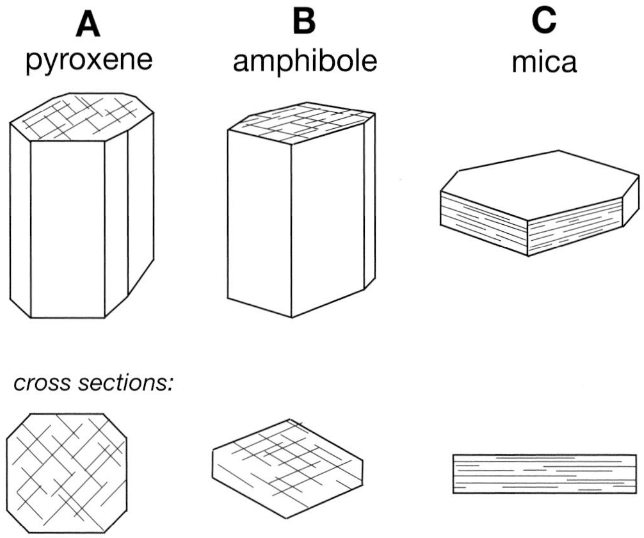

A
pyroxene

B
amphibole

C
mica

cross sections:

Figure A2.7 Common crystal shapes and cleavage planes in pyroxenes, amphiboles and micas. Lower drawings are cross-sections of the upper drawings and show how the cleavage, seen edge-on, might appear in a rock thin section (see Appendix 3).

Pyroxenes have two separate directions of cleavage that intersect at an angle of roughly 90° (Fig. A2.7). The cleavage planes intersect along the direction of Si_2O_6 chains, which are strong. The prefixes 'ortho' and 'clino', incidentally, refer to the so-called crystallographic system, orthorhombic or monoclinic, to which the pyroxene belongs. Orthopyroxene, like all orthorhombic minerals, has a three-dimensional 'building block' of atoms that is brick-shaped. Clinopyroxene's 'building block' (like that of all monoclinic minerals) has four rectangular faces and two parallelogram-shaped faces.

A2.4.3 Double chain silicates

Double chain silicates are also called **amphiboles**. The chains, including the hydroxyl ions associated with them, have a repeat formula of $[Si_4O_{11}(OH)]^{7-}$ (see Fig. A2.6). This is normally doubled to give $[Si_8O_{22}(OH)_2]^{14-}$ simply to give an even number of negative charges, and hence a whole number (7) of positive Ca^{2+}, Fe^{2+}, and Mg^{2+} ions to balance the charge. As with the pyroxenes, the chains are aligned with each other and off-set in two ways to give ortho- and clino-amphiboles. In the **orthoamphiboles**, all the sites between the chains are octahedral (6-fold) and filled with Mg^{2+} and Fe^{2+} ions. The magnesium end-member is called **anthophyllite**, $Mg_7Si_8O_{22}(OH)_2$.

In the **clinoamphiboles**, two of the seven sites are larger (8-fold) and take Ca^{2+} ions, leaving Mg^{2+} and Fe^{2+} ions in the remaining five sites. The simplest formula for a clinoamphibole is the magnesium end-member, $Ca_2Mg_5Si_8O_{22}(OH)_2$, called **tremolite**, which can occur in marble. With substitution of Fe^{2+} it changes to $Ca_2(Mg,Fe)_5Si_8O_{22}(OH)_2$ and is called **actinolite**. By far the most common variety of clinoamphibole is **hornblende**. This has the same formula as actinolite, but with many kinds of limited atomic substitution, notably Na^+ for Ca^{2+}, and Ca^{2+} for Mg^{2+} in the 8-fold site, Al^{3+} for Mg^{2+} in the 6-fod site, and Al^{3+} for Si^{4+} in the four-fold (tetrahedral) site. In addition, there is a space between adjacent chains where large K^+ ions will fit, so hornblende often contains a little potassium. Thus hornblende is unique among common minerals in providing a home for all nine of the main elements found in silicates. A possible formula for hornblende is $K(Ca,Na)_2(Mg,Fe,Al)_5(Si,Al)_8O_{22}(OH)_2$.

A distinctive Na-rich kind of clinoamphibole is called **glaucophane**, $Na_2Al_2(Mg,Fe)_3Si_8O_{22}(OH)_2$. It is a blue mineral, and its Al is in 6-fold sites, indicative of high pressure.

Amphiboles commonly occur as grains with a long, thin shape (described as a prism) with a diamond-shaped cross-section. Two directions of cleavage planes intersect each other at an angle of about 60° and run along the length of the prism (Fig. A2.7).

A2.4.4 Sheet silicates

In the **sheet silicates** the tetrahedra link up to form sheets of hexagonal rings with a repeat formula (including hydroxyl ions) of $[Si_4O_{10}(OH)_2]^{6-}$ (see Fig. A2.5-C). These sheets resemble chicken wire. The tetrahedra sit flat on a plane; the three oxygens lying in the plane are shared, while the fourth oxygen sticks up above the plane (Fig. A2.5-C). A hydroxyl group, $(OH)^-$, is present above the centre of each hexagon; it lies at the same level above the plane as the oxygens that stick up.

Five common kinds of sheet silicate are talc, biotite, muscovite, chlorite and serpentine. A way to visualize their structures is to imagine them as different kinds of sandwich. The 'chicken wire' sheet of linked tetrahedra (called the tetrahedral layer) has the repeat formula $Si_4O_{10}(OH)_2^{6-}$. It can be thought of as a buttered slice of bread, with the buttered side corresponding to the $(OH)^-$ ions and the oxygens that stick up (Fig. A2.5-C).

Talc has the formula, $Mg_3Si_4O_{10}(OH)_2$, and is the simplest sheet silicate. The Mg^{2+} ions sit in octahedral sites between two inward-facing tetrahedral layers, like a slice of Cheddar cheese between buttered slices of bread. The 'sandwiches' are stacked on top of each other with little holding them together, so talc is an incredibly soft mineral; it can easily be rubbed into flaky dust between finger and thumb.

Biotite has a similar cheese-sandwich structure to talc, but one tetrahedron in four contains Al^{3+} instead of Si^{4+} and the charge shortfall is made up by a K^+ ion. The K^+ ions sit between adjacent sandwiches, holding them together. Thus the magnesium end-member of biotite has the formula $KMg_3(Si_3Al)O_{10}(OH)_2$ and biotite, with Fe-Mg substitution, has the formula $K(Mg,Fe)_3(Si_3Al)O_{10}(OH)_2$.

Muscovite is similar to biotite. Both are examples of mica, having K^+ between the sandwiches. Muscovite has Al^{3+} instead of Mg^{2+} and Fe^{2+} as the octahedral 'sandwich filling'. With only two Al^{3+} ions (instead of three Mg^{2+} and Fe^{2+} ions) to provide six positive charges, one in three octahedral sites is vacant. Thus a slice of Emmental cheese, which has large holes, might be a good analogue, instead of Cheddar cheese, for the octahedral aluminium. Muscovite has the formula $KAl_2(Si_3Al)O_{10}(OH)_2$. Al appears twice in the formula to show that it occurs in two different sites.

Chlorite has a structure in which talc-like sandwiches alternate with layers of magnesium hydroxide. Its simplest formula can be written $Mg_3Si_4O_{10}(OH)_2.3(Mg(OH)_2)$, but it always has some Al^{3+} substituting simultaneously for Mg^{2+} in an octahedral site and for Si^{4+} in a tetrahedral site. It also has Fe^{2+} substituting for Mg^{2+}. Thus the general formula for chlorite is $(Mg,Fe,Al)_3(Si,Al)_4O_{10}(OH)_2.3(Mg,Fe,Al)(OH)_2$.

Serpentine has exactly the same formula as the simple magnesium variety of chlorite, i.e. with no Fe or Al. It differs from chlorite in having a structure with 'open sandwiches' in which the 'slices of buttered bread' are all the same way up. Its formula can be written $Mg_6Si_4O_{10}(OH)_8$, or simply as $Mg_3Si_2O_5(OH)_4$.

All the sheet silicates except serpentine are characterized by having perfect cleavage, which is parallel to the sheets (see Fig. 2.10 and Fig. 2.16 in chapter 2). They also commonly form thin, flat, platy crystals which, ideally, have a hexagonal outline (Fig. A2.7) mirroring the shape of the hexagonal rings in the tetrahedral layers. In serpentine, with its open-sandwich structure, the sheets are curved; they can curl and roll up around themselves, resembling sub-microscopic rolled-up carpets, which manifest themselves as serpentine fibres known as white asbestos (see Fig. 2.74).

A2.4.5 Framework silicates

In **framework silicates** the $(SiO_4)^{4-}$ tetrahedra link up to make a strong, three-dimensional scaffolding-like structure. Every oxygen connects two tetrahedra. The simplest example is the mineral quartz (SiO_2), but, because Al^{3+} can substitute for Si^{4+}, framework silicates can also include $(AlO_4)^{5-}$ tetrahedra, as is the case in the mineral **feldspar**, whose end-members were discussed above, namely **K-feldspar**, $KAlSi_3O_8$, **albite**, $NaAlSi_3O_8$, and **anorthite**, $CaAl_2Si_2O_8$. The formulae of feldspars can be remembered as having four lots of SiO_2, giving Si_4O_8, then substituting one or two of the four Si^{4+} ions by Al^{3+} and introducing a K^+, Na^+ or Ca^{2+} ion to balance up the charges. In all cases the large K^+, Na^+ or Ca^{2+} fits into a cage-like space within the framework of tetrahedra. Feldspars between albite and anorthite, which make the plagioclase solid solution series, have separate names for intermediate compositions. Albite and anorthite cover 10% of the range at either end. Thus, albite has up to 10% of anorthite, abbreviated An_{10}, and anorthite has more than 90% anorthite (An_{90}). The rest of the range is divided into four equal 20% intervals with the names oligoclase $(An_{10} – An_{30})$, andesine $(An_{30} – An_{50})$, labradorite $(An_{50} – An_{70})$, and bytownite $(An_{70} – An_{90})$. Plagioclase in metamorphic rocks is usually albite, oligoclase or andesine.

In addition to quartz and feldspar, a third mineral with a framework structure is **cordierite**. Cordierite is an aluminium-rich Fe-Mg silicate whose formula, based on 9 units of SiO_2, is $(Mg,Fe)_2(Si_5Al_4)O_{18}$.

Quartz, feldspar and cordierite all have relatively low densities (and low relief when seen through a microscope – see Appendix 3) because the framework structure prevents the oxygens from being packed tightly together.

A2.5 Minerals in metamorphic rocks

A2.5.1 A list of common minerals

The minerals discussed above are listed in Table A2.1, which shows the formula of each at a glance. Chapter 1 states that 'two dozen or so' minerals comprise most metamorphic rocks. This number includes the silicate minerals reviewed

Table A2.1 Common minerals in metamorphic rocks.

Mineral group and the Si-O-bearing repeat unit with its negative charge	Mineral name and chemical formula
	Indented names are varieties of the main mineral
	Green highlights hydrous silicates
	Blue highlights carbonates

Silicates with independent tetrahedra
$(SiO_4)^{4-}$

Olivine	$(Mg,Fe)_2SiO_4$
Forsterite	Mg_2SiO_4
Fayalite	Fe_2SiO_4
Garnet	$(Fe,Mg,Ca)_3Al_2(SiO_4)_3$
Almandine	$Fe_3Al_2(SiO_4)_3$
Pyrope	$Mg_3Al_2(SiO_4)_3$
Grossular	$Ca_3Al_2(SiO_4)_3$
Kyanite	$Al_2O(SiO_4)$
Andalusite	$Al_2O(SiO_4)$
Sillimanite	$Al_2O(SiO_4)$
Staurolite	$Al_4O_2(SiO_4)_2.FeO(OH)$
Epidote	$Ca_2(Al,Fe^{3+})_3(SiO_4)_3(OH)$

Pyroxenes (single chain silicates)
$(Si_2O_6)^{4-}$

Clinopyroxene	$Ca(Mg,Fe)Si_2O_6$
Diopside	$CaMgSi_2O_6$
Augite	$Ca(Mg,Fe)Si_2O_6$ + substitutions
Jadeite	$NaAlSi_2O_6$
Omphacite	$(Na,Ca)(Al,Mg)Si_2O_6$
Orthopyroxene	$(Mg,Fe)SiO_3$
Enstatite	$MgSiO_3$

Amphiboles (double chain silicates)
$Si_8O_{22}(OH)_2^{14-}$

Clinoamphibole	$Ca_2(Mg,Fe)_5Si_8O_{22}(OH)_2$
Tremolite	$Ca_2Mg_5Si_8O_{22}(OH)_2$
Actinolite	$Ca_2(Mg,Fe)_5Si_8O_{22}(OH)_2$
Hornblende	$Ca_2(Mg,Fe)_5Si_8O_{22}(OH)_2$ + substitutions
Glaucophane	$Na_2Al_2(Mg,Fe)_3Si_8O_{22}(OH)_2$
Orthoamphibole	$(Mg,Fe)_7Si_8O_{22}(OH)_2$
Anthophyllite	$Mg_7Si_8O_{22}(OH)_2$

Sheet silicates
$(Si_4O_{10}(OH)_2)^{6-}$

Talc	$Mg_3Si_4O_{10}(OH)_2$
Muscovite mica	$KAl_2(Si_3Al)O_{10}(OH)_2$
Biotite mica	$K(Mg,Fe)_3(Si_3Al)O_{10}(OH)_2$
Chlorite	$(Mg,Fe,Al)_6(Si,Al)_4O_{10}(OH)_8$
Serpentine	$Mg_6Si_4O_{10}(OH)_8$

Framework silicates
multiples of SiO_2 and $(AlO_2)^-$
$(AlSi_3O_8)^-$

Quartz	SiO_2
Plagioclase	$(Na,Ca)(Si,Al)AlSi_2O_8$
Albite	$NaAlSi_3O_8$
Oligoclase	10-30% anorthite
Andesine	30-50% anorthite
Labradorite	50-70% anorthite
Bytowninte	70-90% anorthite
Anorthite	$CaAl_2Si_2O_8$

$(Al_2Si_2O_8)^{2-}$
$(AlSi_3O_8)^-$
$(Al_4Si_5O_{18})^{4-}$

K feldspar	$KAlSi_3O_8$
Cordierite	$(Mg,Fe)_2Al_4Si_5O_{18}$

Carbonates
$(CO_3)^{2-}$

Calcite	$CaCO_3$
Dolomite	$CaMg(CO_3)_2$
Magnesite	$MgCO_3$

above plus three carbonate minerals, calcite ($CaCO_3$), dolomite ($CaMg(CO_3)_2$ and magnesite ($MgCO_3$), but it does not include the special names used for the end-members or intermediate varieties of olivine, garnet, orthopyroxene, clinopyroxene, clinoamphibole, and plagioclase. Names highlighted in green (amphiboles, sheet silicates, staurolite and epidote) are hydrous. Those highlighted in blue are carbonates. All of them are described in chapter 2, and at the end of that chapter (in Fig. 2.78) their compositions are shown on a so-called ACF triangular diagram.

A2.5.2 Accessory minerals and minerals in unusual kinds of rock

In addition to the minerals named above, there are ten others that are often present in tiny amounts in metamorphic rocks; these are called **accessory minerals**. Four of them are iron-rich, namely magnetite (Fe_3O_4), hematite (Fe_2O_3), ilmenite ($FeTiO_3$) and pyrite (FeS_2). These four are often referred to as 'opaques' because they appear as black silhouettes when seen in a thin section through a microscope (see Appendix 3). The other six each contain an element that is not one of the ten elements in Figure A2.1. The element titanium (Ti)

occurs in rutile (TiO_2) and titanite ($CaTiSiO_5$), phosphorus (P) occurs in **apatite** ($Ca_3(PO_4)_3(OH,F,Cl)$), zirconium (Zr) occurs in zircon ($ZrSiO_4$), chromium (Cr) is present in chrome-spinel (chromite) whose formula is $(Fe,Mg)Cr_2O_4$, and boron (B) occurs in **tourmaline**, which is a ferromagnesian silicate with a complex, variable formula.

Finally, a further 14 minerals are named in the text, mostly in chapter 4, and are described only briefly. They mainly occur in rocks that formed under extreme metamorphic P–T conditions – very low temperature, high temperature at very low pressure, and very high to extremely high pressure. These fourteen minerals are zeolite (a group of hydrous framework silicates), three Ca-Al-rich silicates called prehnite, pumpellyite and lawsonite, the Ca silicate mineral wollastonite, the Mg oxide and Mg hydroxide minerals called periclase and brucite respectively, the Al oxide mineral called corundum or sapphire, the high-pressure polymorphs of SiO_2 called coesite and stishovite, the polymorphs of carbon (graphite and diamond), the high-pressure polymorph of $CaCO_3$ called aragonite, and the high-pressure polymorph of $(Mg,Fe)_2SiO_4$ (olivine's composition) called ringwoodite.

Appendix 3   Minerals under the microscope

The standard way of examining rocks in close detail is to prepare them as **thin sections** and to look at them through a **polarizing microscope**. Anyone with access to the internet can study thin sections of rock at their leisure using the **Virtual Microscope**. This is an Open University website (virtualmicroscope.org) that provides free access to images of several hundred different rocks. Users can zoom in and out, move around the thin section, and adjust the viewing conditions almost as if they were looking through a real microscope. Many of the images of thin sections in this book are taken from the Virtual Microscope, and most of the metamorphic rocks in the Irish University Rocks collection (the GeoLab collection) were commissioned with the book in mind.

This appendix explains how thin sections are made; it describes the operation of a polarizing microscope and the features of minerals that can be seen when using it, and it suggests a practical approach to identifying common minerals in a thin section.

A3.1 Thin sections

A thin section is a slice of rock, just 30 microns (0.03mm) thick, cemented to a glass microscope slide with epoxy resin. Most minerals at this thickness are translucent (they let light through), so they can be examined using a microscope with the illumination from below.

The preparation of thin sections requires high-precision machinery and a dedicated workshop (Fig. A3.1). First, using a water-cooled diamond-tipped circular saw, a slice of rock a few millimetres in thickness is cut from a rock specimen. A rectangular tablet, typically about 2cm by 4cm, is trimmed from the slice of rock, and one face of it is ground perfectly flat on a lapping wheel. The flattened face is then cemented firmly to a glass microscope slide using clear epoxy resin. The bulk of the tablet is skimmed off with a thin-bladed circular saw leaving a veneer of rock, perhaps 200 microns (0.2mm) thick, attached to the glass slide. The veneer of rock is then ground away gently on a lapping wheel until a thickness of just 30

Figure A3.1 Equipment used, and one of the intermediate steps, in preparation of thin sections. Top left: water-cooled rock saw. Top right: rectangular rock tablet glued to a glass microscope slide. Bottom left: precision circular saw for skimming off most of the tablet. Bottom right: grinding wheel for the high-precision lapping of the skimmed rock, down to 30 microns.

microns remains. The thin section is now almost finished. Final preparation is in one of two ways. The traditional way is to make a covered thin section. Here, a hot, sticky coating of a natural resin called Canada balsam is smeared over the surface and a glass cover slip is gently pressed onto the resin, avoiding entrapment of air bubbles (Fig. A3.2). The covered section is then baked to cure the resin. The second way is to make a polished thin section. In this case the surface is not covered, but simply polished to a mirror finish using gamma alumina or diamond paste on a polishing lap.

Covered thin sections give the best images under the microscope because they show a feature called **relief**. Relief is the apparent roughness of the surface of a mineral grain. Minerals that look flat and featureless, like quartz, are said to have low relief, whereas those that stand out with bold outlines and a rugged surface, like garnet, are said to have high relief (Fig. A3.3).

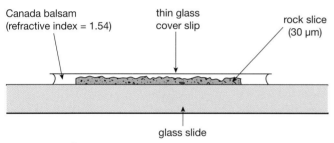

Canada balsam
(refractive index = 1.54)

thin glass
cover slip

rock slice
(30 μm)

glass slide

Figure A3.2 Cross-section through a standard covered thin section.

Figure A3.3 Contrasting examples of relief. This microscope image (a screen shot of the Virtual Microscope rock, GeoLab M06) is 2mm wide and shows three grains of garnet surrounded by quartz in a covered thin section. Garnet has high relief: it looks greyish and roughened, with bold margins and dark cracks. Quartz has low relief: its surface is so flat and featureless it can hardly be seen.

Relief depends on an important property of a mineral called **refractive index**. This is the ratio of the speed of light in a vacuum to its (lower) speed through the mineral, so it is a number greater than unity. Minerals have low relief if their refractive index is close to that of Canada balsam, which is 1.54. The refractive index of quartz is about 1.55 and so quartz has low relief. In contrast, garnet's refractive index is about 1.8, so garnet has high relief.

Polished thin sections, on the other hand, can be examined in several different ways. As well as being visible under a regular polarizing microscope, they can be looked at using a reflected light microscope, and they are also amenable to study using an instrument called a **scanning electron microscope (SEM)**. An SEM produces wonderful images, and also permits the *in situ* chemical analysis of tiny parts of mineral grains. The technique is described in Appendix 4.

The method for preparing thin sections was developed, incidentally, in the mid-nineteenth century by Henry Clifton Sorby (Fig. A3.4). Sorby was a brilliant and independently wealthy geologist, biologist, metallurgist, and microscopist who was elected as a Fellow of the Royal Society at the age of 31. Sorby's method of making thin sections revolutionized the

Figure A3.4 Henry Clifton Sorby (1826–1908) who, amongst his many scientific achievements, invented the process for making thin sections of rock. *Source: Wikipedia.*

study of rocks because it allowed the relationships between grains to be examined under a microscope for the first time. Until then, only loose, crushed mineral grains could be seen with a microscope.

A3.2 The polarizing microscope

The polarizing microscope is a normal transmitted-light microscope, as used in biology, but with two modifications. Firstly, the microscope stage is circular and can be rotated. Secondly, two polarizing filters (made from similar material to that used in polarizing sun glasses) are positioned in the light path, one below the stage and the other above the stage (Fig. A3.5). One of the filters is fixed permanently in place. This is usually the filter below the stage. The other filter is removable, and can be moved easily into the light path and out again. Each filter has a so-called polarization direction (also called the privileged direction, or the vibration direction). A beam of light shining from the light source up through the lower filter will emerge from that filter with its light waves vibrating parallel to the filter's polarization direction, and the light is then said to be **plane polarized light**. The polarization direction of the filter below the stage is at right angles to that of the filter above the stage (Fig. 3.5, right-hand side); the two directions are said to be 'crossed'.

With no thin section on the stage, an observer will see just blackness. The polarized light from the lower filter is completely absorbed by the 'crossed' upper filter. However, when a thin section is placed on the stage, its mineral grains become clearly visible, as if by magic. They appear in shades of grey, white and various colours. Thin sections viewed in this way are said to be viewed between **crossed polars** (shortened from 'crossed polarizing filters' and abbreviated **XP**).

This remarkable behaviour can be demonstrated using a light table and two squares cut from a sheet of Polaroid plastic film (Fig. A3.6). The polarization directions of the two squares, marked by ruled black lines on their edges, are 'crossed', just

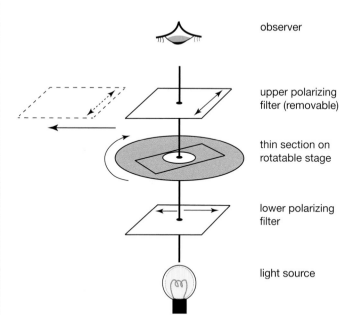

Figure A3.5 Left: a polarizing microscope (a Leitz Pol student model, more than 50 years old and still working well). Right: diagram showing how light passes vertically up the microscope. The double-ended arrow on each polarizing filter shows its polarization direction.

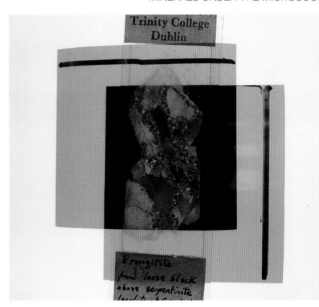

Figure A3.6 Two partly overlapping squares of Polaroid plastic film (polarizing filter) on a light table. The filters are 'crossed', i.e. the polarization direction (the black line) of one filter is at right angles to that of the other filter. The result is that no light passes through the area of overlap. On the right, a thin section is sandwiched between the two filters. The mineral grains allow light through, displaying different colours called interference colours.

like the crossed polars in a microscope. On the left in Figure A3.6, where the two squares overlap, no light gets through. On the right, however, where a thin section has been slipped between the Polaroid squares, the rock can be seen clearly, each grain displaying a different colour. These colours are produced in the upper polarizing filter by a physical process known as interference of light, so they are called **interference colours**.

Interference colours depend on the value of a mineral property called **birefringence** (pronounced by-ra-**fringe**-nss). The correspondence between interference colours and birefringence values for a standard thin section (0.03mm thick) is shown in the top part of Figure A3.7. Here, birefringence values range from zero up to 0.06, and the related interference colours are divided into so-called **orders**. The first order has grey, white, yellow, orange and deep red interference colours, changing gradually from one to the next. The second order has bright purple, blue, green, yellow, orange and deep red colours. The third order is dominated first by green and then by pink. Higher orders are not shown.

Eleven different minerals are named on the lower half of Figure A3.7. Any particular one of them, with the exception of garnet, will show interference colours that vary, from one grain to another in a thin section, and correspond to birefringence values between zero and some maximum (indicated by the dashed line in Figure A3.7 that ends with the mineral's name). The maximum birefringence value for a mineral is given the symbol Δ and is an important property of that mineral. For quartz ($\Delta = 0.009$) the interference colours will only be grey and white, whereas olivine ($\Delta = 0.04$) will display various colours, including all the first-order colours, and second-order colours up to bright orange, going from one grain to another. The mineral calcite has $\Delta = 0.17$ so it is well beyond the range covered in Figure A3.7. Its interference colours are mostly creamy white, called high-order white. Garnet, the exception, only ever appears black between crossed polars (XP) and is said to be **isotropic**.

Clearly, the interference colour from a single grain is of little significance for identifying the mineral; what is important is the full range of different interference colours seen in

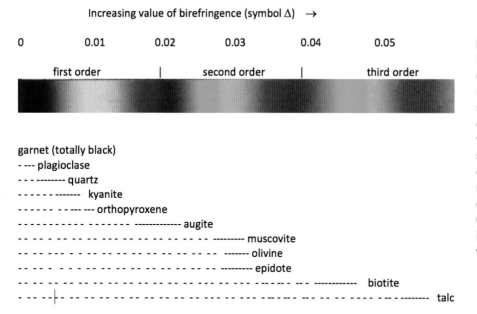

Increasing value of birefringence (symbol Δ) →

0   0.01   0.02   0.03   0.04   0.05

first order | second order | third order

garnet (totally black)
- - - plagioclase
- - - - - - - - - quartz
- - - - - - - - - - - - kyanite
- - - - - - - - - - - - - orthopyroxene
- augite
- muscovite
- olivine
- epidote
- biotite
- talc

Figure A3.7 Diagram summarizing birefringence and interference colours of common minerals. The upper part shows so-called first, second and third-order interference colours and how they relate to the value of birefringence. Below is shown the range of interference colours normally seen in XP in randomly orientated grains of eleven common minerals in a standard 0.03mm thin section. The range is indicated by the dashed line ending with the mineral's name.

many grains. This allows Δ to be estimated and Δ is a helpful parameter for identifying a mineral.

But what is birefringence? Birefringence is related to refractive index. The refractive index of a mineral is not fixed at a single value, but has a *narrow range* of values. This *range* of refractive index is the mineral's birefringence. The range for quartz, for example, is between 1.545 and 1.553, so the birefringence of quartz is 1.553 minus 1.545, or 0.009. The refractive index value of 1.55 for quartz, stated in the previous section in the context of relief, is only the *average* value for quartz. The precise value of refractive index for quartz varies from grain to grain in a thin section, and depends on the *direction* the light is travelling through it relative to its internal crystal structure. Light travels more slowly in some directions (so the mineral's refractive index is higher) than it travels in other directions. Since refractive index depends on the orientation of a grain, so too does birefringence.

Returning to the polarizing microscope, when the removable filter is retracted, so that only one filter is in the light path, a thin section is said to be viewed in **plane polarized light** (abbreviated **PPL**). In PPL the 'true' colour of a mineral (as opposed to its interference colour) is seen. This colour is produced when some of the colours from the visible spectrum of light are absorbed by the mineral grain as the light passes through it. The colour of the emerging light is complementary to what is absorbed. It is sometimes called the **absorption colour** of the mineral. As an example, the mineral chlorite has a green absorption colour. It appears green because it absorbs all colours of the spectrum other than green, i.e., it absorbs red, orange, yellow, blue and purple light. Only green light is allowed through. Many minerals have no colour and are described as colourless. A few absorb all the light, letting none through. They appear as black silhouettes in PPL, and are described as opaque minerals or simply opaques.

When the microscope stage is rotated, significant changes in the appearance of the grains can be observed in both XP and PPL.

In XP the interference colour of each mineral grain keeps the same hue, but becomes gradually brighter and darker. At its darkest the grain is totally black, and this state of blackness is termed **extinction**. Extinction happens four times when the microscope stage is turned through a complete revolution of 360°. In other words, a mineral grain will go into extinction once in each 90° of rotation. Extinction happens because each mineral grain behaves like a small piece of Polaroid film. However, unlike Polaroid, the grain has not one, but two, polarization directions. These two directions are at 90° to each other, and extinction happens when rotation of the stage brings them into alignment with the polarization directions in the microscope's two filters (Fig. A3.8), which are described as being oriented, respectively, east–west (E–W) and north–south (N–S).

Many minerals whose grains have elongate shapes in thin section go into extinction when the length of the grain is parallel to the N–S or to the E–W polarization direction of the filters, i.e. parallel to one of the cross-hairs in the microscope. This is called **parallel extinction**, or **straight extinction**. It happens, for example, with muscovite and biotite (see Fig. A3.9). Other minerals with elongated grains go into extinction when the length of the grain is inclined to the cross hairs. This is called **inclined extinction**. It happens, for example, in hornblende where the angle between the grain length and the nearest cross-hair is quite small, generally between

0° and about 20°. The angle of extinction in some minerals is variable up to 45° (the maximum).

What happens in PPL? In PPL as the microscope stage is rotated the absorption colour of a mineral grain will generally change. The change is termed **pleochroism** (pronounced plee-oh-**croh**-ism). The minerals biotite, hornblende and chlorite, for example, all show pleochroism. They are called pleochroic minerals. In each of them, the colour changes to a new colour (or a different depth of the same colour) following a rotation of 90°, and returns to the initial colour if the stage is turned through a further 90°. The pleochroism of biotite can be examined online (virtualmicroscope.org) in VM GeoLab section M09 at Rotation 1, which is shown in Figure A3.9. The biotite forms the elongate grains that are deep reddish brown when aligned with the east–west cross-hair, and change to pale golden brown when aligned with the north–south cross hair.

Six minerals, including biotite, can be seen at rotation 1 in M09 (Fig. A3.9) and their distinctive properties under the microscope are now described. To make sense of the descriptions, the reader should now go online and look carefully at M09, in PPL and XP, as it is rotated.

Biotite, as noted, is brown and pleochroic in PPL. The pleochroism is more pronounced in some grains than others, which is always the case with pleochroic minerals. The obvious dark circular patch on one grain is a so-called pleochroic halo – a zone where radiation damage has made the biotite intensely pleochroic. The radiation came from a radioactive element, perhaps uranium, in a tiny grain of the accessory mineral, zircon, trapped inside the biotite. The zircon is just about visible at the centre of the halo. In XP

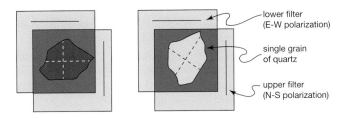

Figure A3.8 Diagram illustrating extinction in a single mineral grain of quartz in XP. The mineral acts like a double polarizing filter, with two polarization directions at right angles, marked by dashed lines. On the left, the grain is in extinction because the polarization directions of the grain are aligned with those of the microscope's filters. On the right, after rotating the stage, the grain is no longer in extinction and light gets through.

lower filter (E-W polarization)

single grain of quartz

upper filter (N-S polarization)

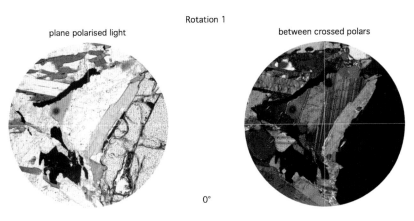

Rotation 1

plane polarised light

between crossed polars

0°

1mm

Figure A3.9 Screen shot of rotation 1 in thin section M09 (garnet-mica schist) from the GeoLab collection at the Virtual Microscope (virtualmicroscope.org). Six minerals are present, and their behaviour as the stage is turned is described in the text.

the biotite shows green and pink (third-order) interference colours, and parallel extinction.

Muscovite is seen as the elongate, parallel-sided, colourless grain in the top left quadrant. In XP it shows greeny-yellow (second-order) interference colours, and it has parallel extinction like biotite. As the stage is turned and extinction is approached, muscovite takes on a speckled appearance with bright flecks decorating its darkening surface. This speckled look is a distinctive feature of all kinds of mica; it can also be seen in biotite.

Plagioclase forms the large colourless grain with low relief covering the centre of the field. In XP its grey (low first-order) interference colours appear as parallel stripes. The stripes are due to so-called lamellar twinning; each grain is made of many parallel plate-like sub-crystals, known as **twin lamellae**. The stripes have inclined extinction. The angle of inclination here is small, which is a feature of Na-rich plagioclase (see Appendix 2).

Quartz, like plagioclase, is colourless with low relief. Only a few small grains are present in this field of view. One of them sits at the top end of the large grain of plagioclase. In XP it has no stripes, and it stands out here because its interference colour is white, rather than grey.

Garnet is the very large grain with high relief on the right. Only part of it is visible. In XP it is totally black because its birefringence is zero; it is optically isotropic.

Opaque grains constitute the sixth mineral (or minerals). They are seen as black silhouettes in PPL, and they are obviously also black in XP. Common opaque minerals are magnetite and hematite (both are oxides of iron), and pyrite (iron sulphide) but they cannot be distinguished using a polarizing microscope.

A hole in the slide might be mistaken for a seventh 'mineral'. It is colourless with low relief, and it is isotropic. The example here is completely surrounded by garnet. This 'mineral' is a place where part of the garnet grain was plucked out and lost when the thin section was being made. Holes are quite common in thin sections.

Finally, a diagnostic feature of some minerals in PPL, and not mentioned so far, is **cleavage**. This is the predisposition for a mineral grain to split along perfectly flat internal planes, called cleavage planes. Cleavage in muscovite, for example, is illustrated in chapter 2 (Fig. 2.10). When a thin section is being made, stresses induced by sawing and grinding may cause cleavage planes to crack and gape very slightly, making them visible as perfectly straight, parallel lines running across the grain in thin section. Unfortunately, with modern lapping machines, grinding is not aggressive (as once it was) and cleavage cracks do not always open up, except perhaps around the margins of a thin section where stresses during manufacture would have been greatest. Cleavage lines can be used to say whether a mineral has parallel or inclined extinction. Also, some minerals have more than one set of cleavage planes, which can aid their identification.

A3.3 Identifying minerals

Since there are a limited number of common minerals, most of these can usually be identified successfully by a simple process of elimination using some kind of flow diagram. The approach suggested here is to use a chart where minerals are placed in one of 16 boxes in a four-by-four grid, as in Figure A3.10. The four rows are based on interference colours in XP. From top down, they correspond to minerals with interference colours up to 1) high-order creamy white, 2) up to purple, blue and other second and third order colours, 3) up to grey, white, cream, yellow and deep orange of the first order, and 4) black (isotropic minerals).

The four columns correspond to features seen in PPL. Colourless or weakly coloured minerals go into the first three columns. The first column is reserved for those with low relief, like quartz. The second column is for minerals with high relief, but without cleavage, like garnet. The third column is for minerals with (generally) high relief, and which have cleavage. Strongly coloured minerals are in the fourth column.

A mineral whose identity is sought is first placed in its box by deciding its range of interference colours and its relevant properties in PPL. Identification then depends on matching its properties with those of one of the minerals named in that box, and to eliminate the others. To help here, small drawings in each box show distinctive properties such as relief, cleavage and colours in PPL, and twinning in XP. Also the drawings are orientated to show at a glance whether extinction is parallel (e.g. muscovite), slightly inclined (e.g. hornblende), or highly inclined (e.g. clinopyroxene). Further distinguishing features are described, box by box, in the paragraphs below.

The two-dozen or so main minerals listed in Appendix 2 are all present in the diagram, though only some of the special names for end-members and intermediate compositions are

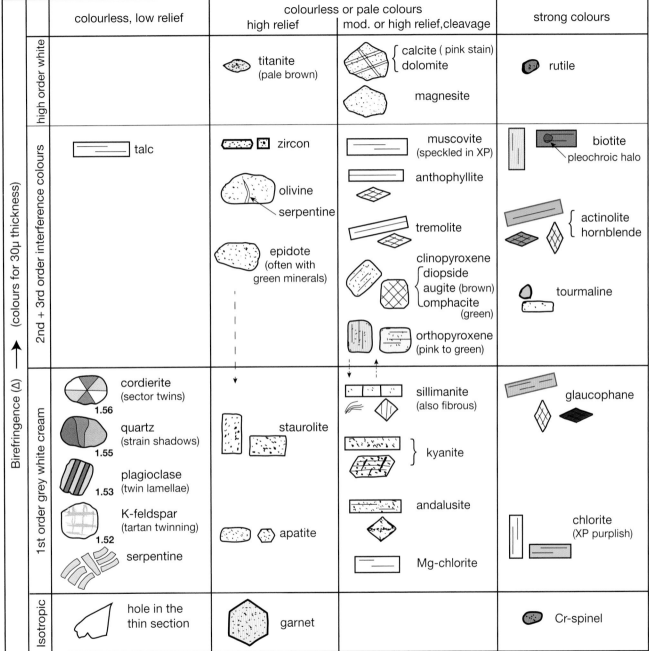

Figure A3.10 A chart to help with the identification of minerals seen in thin section through a polarizing microscope. For an explanation of its use, and of how to distinguish between minerals in the same box, see the text.

given. Six of the accessory minerals named in Appendix 2 (those that are not opaque) are also included, and are distinguished from the main minerals by being shown as *smaller* drawings.

In practice, one tends to recognize minerals through their association with other minerals in particular kinds of rock. It is then just a question of confirming a mineral's identity with a few quick checks. Almost all of the minerals named on the diagram can be seen in at least one, and usually several, of the thin sections illustrated in the book as recorded in Table A3.1. Exceptions are the clinoamphiboles tremolite and actinolite, and the accessory mineral zircon.

Column 1, row 3 Cordierite, quartz, Na-plagioclase, Ca-plagioclase, K-feldspar, and serpentine (low relief and colourless in PPL, with only grey and white colours in XP)

These six are more easily distinguished in XP. Plagioclase may have parallel grey stripes (twin lamellae). The angle of extinction is always less than 20° for Na-plagioclase but can be much higher for Ca-plagioclase. A distinctive feature of quartz is **undulose extinction**, or **strain shadowing**, where a grain of quartz consists of several sub-grains that go into extinction separately as the stage is rotated through a few degrees. A good example can be seen in VM GeoLab section M26 (eclogite) at rotation 2. K-feldspar in XP commonly has criss-crossing grey stripes, called **tartan twinning**. K-feldspar can be distinguished from accompanying quartz in PPL by lowering the microscope stage to defocus very slightly. A bright edge, fringing K-feldspar, shifts into adjacent quartz, leaving the K-feldspar dimmer. This procedure is called the **Becke test**, and an example of it is shown in chapter 5, Figure 5.13. The bright fringe, called the Becke line, always moves towards the mineral with the higher refractive index. Refractive index values for cordierite, quartz, Na-plagioclase and K-feldspar are written on the chart (Fig. A3.10). Quartz can also be distinguished by its so-called interference pattern, which appears as a dark cross, but an account of the method for observing interference patterns is beyond the scope of this book. Cordierite often accompanies andalusite; it may be riddled with dusty inclusions, and can be twinned in wedge-shaped sectors. Serpentine is a major alteration product of olivine, and has a distinctive interlocking pattern of grey, wavy bands in XP.

Column 2, row 2 Olivine and epidote (colourless or faintly coloured in PPL, high relief, no cleavage, up to second or third order interference colours)

One way to tell olivine from epidote is by association. Epidote, being hydrous, is commonly associated with coloured hydrous minerals such as hornblende, chlorite or glaucophane. Olivine, being anhydrous, is not normally found alongside these minerals, but it is commonly in the process of being altered to serpentine. Epidote may be yellow in PPL, unlike olivine, and in XP epidote's interference colours are often quite variable within a single grain due to variation in the content of Fe^{3+}. In fact, epidote with no Fe^{3+} has only first-order grey colours and appears in row 3, hence the vertical dashed arrow straddling two boxes.

Column 3, row 1 Calcite, dolomite and magnesite (high relief, colourless and 'high-order white' interference colours)

Calcite and dolomite in XP may both have one, two or even three sets of distinctive criss-crossing coloured stripes, which are twin lamellae. The two carbonates are so similar in appearance that in order to tell them apart the thin section is usually stained with Alizarin-red dye before the cover slip is applied. The dye makes calcite pink but does not affect dolomite. If a thin section has not been so stained, and calcite or dolomite is suspected, it is good practice to use the generic name carbonate, rather than to guess. The third carbonate, magnesite, never has twin lamellae. The high-order white interference colour of carbonates can readily be distinguished from first-order whites and greys by using a special filter, called a sensitive tint plate, which is inserted above the thin section in the microscope. It causes first-order white to change to either yellow or blue, but leaves high-order white unaffected. This test, unfortunately, is not available on the Virtual Microscope.

Column 3, row 2 Muscovite, anthophyllite, tremolite, clinopyroxene (diopside, augite, omphacite), and orthopyroxene (colourless or faintly coloured with high relief and cleavage; up to bright second-order colours in XP)

Muscovite is distinctive. It is commonly elongate in outline with moderate relief and parallel extinction, and a speckled appearance in XP when it is close to extinction (as described above). Anthophyllite similarly has parallel extinction, but is not speckled in XP. Tremolite resembles anthophyllite but with slightly inclined extinction. End-on it may show two cleavage directions about 60° apart, like all amphiboles. Clinopyroxene's extinction is highly inclined to the cleavage while orthopyroxene has parallel extinction. Both pyroxenes may be faintly coloured and show 90° intersecting cleavage. Orthopyroxene rarely has interference colours beyond the first order, so a dashed arrow shows it can occur also in row 3. In PPL it may show pleochroism from pale pink to pale green.

Columns 2 and 3, row 3 Andalusite, sillimanite, kyanite, staurolite (high relief, elongate outlines, up to first-order grey, white, or yellow interference colours)
All four minerals occur in metapelite. Staurolite (in column 2) in PPL is yellow and pleochroic with no cleavage and a rectangular outline. The others have prismatic shapes with parallel extinction, and can usually be distinguished when seen end on. Andalusite has a square outline, often with a dark, diagonal cross-shaped pattern of inclusions. Sillimanite also has square outlines, but with a single diagonal cleavage. In XP its interference colours may be up to second order, so an arrow shows that it may occur in row 2. Kyanite has one good cleavage and a less good cleavage at about 100° to it. It is prone to being bent slightly.

Column 4 Biotite, hornblende, actinolite, glaucophane, and chlorite (strongly coloured minerals)
Glaucophane has obvious blue-purple-colourless pleochroism. The other four minerals in PPL are green or brown, and pleochroic, and can be confused. Biotite has parallel extinction and it looks speckled when close to extinction. Hornblende has slightly inclined extinction, typically 10 to 15°, and its interference colours are up to second-order blue (not green or pink like biotite). End-on sections may show two cleavages that intersect at about 60°. Actinolite cannot readily be distinguished from pale green hornblende. Chlorite with elongate outlines has parallel extinction, and usually has so-called anomalous first-order dark purple or steel-blue interference colours. If chlorite is magnesium-rich, it will be pale green or even colourless, so the mineral name appears also in column 3.

The six **accessory minerals** are distinctive enough. In column 2, row 1, *titanite* has high-order white interference colours. In PPL it is pale brownish and has very high relief. *Zircon* (column 2, row 2) is like olivine but forms tiny prisms with a square cross-section and *very* high relief. *Apatite* (column 2, row 3) forms colourless high-relief rounded grains that go dark grey, or almost black, in XP. An example is shown in Figure A3.11. The last three accessory minerals are all strongly coloured in PPL, so appear in column 4. *Tourmaline*

Figure A3.11 Two rounded grains of the accessory mineral apatite in the garnet-mica schist, M09. In PPL it is colourless, with high relief and without cleavage; in XP it has dark grey to black interference colours. Like other accessory minerals in the identification chart (Fig. A3.10) apatite is widespread, but rarely abundant. The other minerals seen here are also present in Figure A3.9. Can you name them?

is prismatic and pleochroic, with a cross-section shaped like a bulging triangle. Elongate grains are deeply coloured when aligned N–S, in contrast to elongate grains of biotite and hornblende, which are deeply coloured when aligned E–W. *Chrome spinel* is brown and isotropic. *Rutile* is brown, and is not isotropic.

To conclude this section on identifying minerals, Table A3.1 has been compiled as a guide to where in the book the different minerals can be seen, and where many of them can be examined at leisure using the Virtual Microscope. The table is in two parts. The first part is a numbered list, 1 to 41, of figures with images of thin sections. Each entry gives the Virtual Microscope (VM) collection and number (if relevant), the name of the rock, the place where it was collected, the rock's approximate *metamorphic* age, and its mineral assemblage. Many of the rocks are from the Grampian orogenic belt and are shown as having formed 470Myr ago.

The second part is a list of minerals, alphabetically under headings of main minerals, accessory minerals, and minerals not included in the chart (Fig. A3.10). Beside the name of each mineral are the number(s) of the rocks, from the first list, in which that mineral can be seen. It should be possible, therefore, to pick any mineral and quickly find examples of what it looks like, or, alternatively, to pick a VM rock and see if its minerals can be recognized before checking against the mineral assemblage in Table A3.1.

Table A3.1 Rocks and minerals shown here as thin sections.

Part 1: List of figures that are images of thin sections.

1) Fig. A3.9, Fig. A3.11; VM GeoLab M09; Garnet mica schist; from Glenshiel, Inverness-shire, NW Scotland; 470 Myr; contains garnet (augen), plagioclase, quartz, muscovite, biotite, opaques and apatite.

2) Fig. A3.3, Fig. 2.6, Fig 2.7, Fig. 5.6; VM GeoLab M06; Garnetiferous quartzite; from Slishwood, Sligo, NW Ireland; 470 Myr or earlier; contains quartz (much as ribbons), with a little garnet, and K-feldspar or perthite.

3) Fig. 2.21; VM GeoLab M08; Garnet-mica-schist; from Coolaney, Sligo, NW Ireland; 470 Myr; contains garnet (late-grown poikiloblasts), quartz, muscovite, and chlorite.

4) Fig. 2.22, 2.25; VM GeoLab M10; Garnet-mica-schist; from Connemara, W Ireland; contains garnet, muscovite, biotite, staurolite (late-grown), quartz, and a trace of tourmaline,

5) Fig. 2.23; VM Open University S339-24; Kyanite schist; from the Scottish Highlands; 470 Myr; contains kyanite, garnet, quartz, muscovite, biotite.

6) Fig. 2.24; not VM; Sillimanite rock; location and age unknown; contains sillimanite.

7) Fig. 2.31; VM Leeds 14; named 'slate' but actually andalusite-cordierite hornfels; from the English Lake District; 400 Myr; contains andalusite, cordierite, biotite, quartz.

8) Fig. 2.38, Fig. 2.39; VM GeoLab M05; Forsterite marble; from Glenelg, NW Scotland; 1000 Myr; contains calcite (stained pink), dolomite, forsterite, diopside, some mica, and secondary serpentine.

9) Fig. 2.46; VM GeoLab M15; Greenschist; from Horn Head, Donegal, NW Ireland; 470 Myr; contains green chlorite, epidote, quartz and albite (not distinguishable), biotite.

10) Fig. 2.48; VM GeoLab M02; Epidote amphibolite; from Coolaney, Sligo, NW Ireland; 470 Myr; contains green hornblende, epidote, untwinned albite, quartz, titanite, rare biotite.

11) Fig. 2.50; not VM; Amphibolite; from Glenelg, Inverness-shire, Scotland; 470 Myr; contains green hornblende, plagioclase, garnet.

12) Fig 2.51, Fig 3.19; VM GeoLab M24; Two pyroxene granulite; from Scourie, Sutherland, Scotland; 2400 Myr; contains clinopyroxene (augite), orthopyroxene, plagioclase.

13) Fig. 2.53; VM GeoLab M12; Glaucophane schist; from Ile de Groix, Brittany, France; 300 Myr; contains glaucophane, epidote, garnet.

14) Fig. 2.55, 2.56 and Fig. 3.25; VM GeoLab M26; Eclogite; from Glenelg, Inverness-shire, Scotland; 1000 Myr; contains omphacite, garnet, rutile, quartz, secondary plagioclase and hornblende..

15) Fig. 2.63; VM GeoLab M01; Amphibolite; from Lough Nagilly, Donegal, NW Ireland; 400 Myr; contains green-brown hornblende, plagioclase, opaques.

16) Fig. 2.67; VM Open University S339-11; Quartz-feldspar mylonite; Assynt, Sutherland, Scotland; 470 Myr; contains strained quartz, feldspar.

17) Fig. 2.68; not VM; Mylonite with pseudotachylite; from Glenelg, Inverness-shire, Scotland; 470 Myr; contains quartz, K-feldspar.

18) Fig. 2.70; VM GeoLab M19; Spinel lherzolite; from SE Spain; recent; contains olivine, orthopyroxene, clinopyroxene, chrome spinel, late plagioclase.

19) Fig. 2.71; VM GeoLab M25; Dunite; from Almklovdal, W Norway; 400 Myr; contains olivine, and a little orthopyroxene (enstatite), chlorite, talc.

20) Fig. 2.73; VM GeoLab M18; Serpentinite; from Slishwood, Sligo, NW Ireland; 470 Myr; contains serpentine, chrome spinel, opaques.

21) Fig. 2.76; VM GeoLab M27; Talc-magnesite rock; from Westport, W Ireland; 470 Myr; contains talc, magnesite, some dolomite, opaques.

22) Fig. 2.77; VM GeoLab M03; Anthophyllite-magnesite rock; from Scourie, Sutherland, Scotland; 1700 Myr; contains anthophyllite, magnesite, opaques.

23) Fig. 3.10; not VM; Dolerite; from the Scourie Dyke, Sutherland, Scotland; 2400 Myr; contains augite, plagioclase.

24) Fig 3.19; VM UK collection; Pyroxene hornfels; from Isle of Rum, Scotland; 60 Myr; contains pyroxene, Ca-plagioclase, opaques.

25) Fig. 3.24; not VM; Cataclastic pyroxenite; from Glenelg, Inverness-shire, Scotland; 470 Myr; contains strained clinopyroxene.

26) Fig. 4.3; VM UK collection; Brucite marble; from Skye, NW Scotland; 60 Myr; contains calcite, brucite.

27) Fig. 4.5; not VM; Pelitic hornfels; from Comrie, Scottish Highlands; 400 Myr; contains corundum (sapphire), biotite.

28) Fig. 4.6; VM GeoLab M20; Andalusite hornfels; from Ardara, Donegal, NW Ireland; 400 Myr; contains andalusite porphyroblasts, muscovite, biotite, quartz.

29) Fig. 4.15; VM GeoLab M13; Glaucophane schist; from Achill Island, Mayo, W Ireland; 470 Myr; contains glaucophane, untwinned albite poikiloblasts, tiny garnet, muscovite, chlorite.

30) Fig. 4.21; not VM; Coesite; location and age unknown; contains coesite, garnet, secondary quartz.

31) Fig. 4.26; VM UK collection; Triassic meteorite impact deposit; from near Bristol, England; late Triassic 220 Myr; contains fine-grained minerals including chlorite.

32) Fig. 4.27; not VM; Anorthosite; from the Apollo 16 site, the Moon; 4400 Myr; contains Ca-plagioclase (anorthite).

33) Fig. 4.28; not VM; Shock melted anorthosite; from the Apollo 17 site, the Moon; 4000 Myr; contains Ca-plagioclase, diaplectic glass.

34) Fig. 4.29; not VM; Shock melt vein; from the Taiban meteorite; 470 Myr (shock age); contains ringwoodite, olivine.

35) Fig. 5.3; VM GeoLab M07; Garnet-kyanite-gneiss; from Slishwood, Sligo, W Ireland; 470 Myr or older; contains strained kyanite, garnet, quartz (some ribbons), albite, K-feldspar, biotite.

36) Fig. 5.4; not VM but from same rock as VM GeoLab M08; Mesoperthite; from Slishwood, Sligo, NW Ireland; 470 Myr or older; mesoperthite is an even mixture of albite and K-feldspar

37) Fig. 5.7; VM GeoLab M23; Garnet-clinopyroxene-plagioclase granulite; from Slishwood, Sligo, W Ireland; 470 Myr or older; contains garnet, clinopyroxene, plagioclase, hornblende, quartz, rutile.

38) Fig. 5.11; GeoLab M22; Pelitic gneiss; from Slishwood, Sligo, W Ireland; 470 Myr or older; contains quartz, kyanite replacing sillimanite, muscovite, biotite.

39) Fig. 5.12; not VM; Symplectite after omphacite; from Glenelg, Inverness-shire, Scotland; 1000 Myr; contains omphacite changing to intergrown albite and hornblende.

40) Fig. 5.13; not VM; Psammitic gneiss; from Slishwood, Sligo, W Ireland; 470 Myr or older; contains kyanite, quartz and a reaction border of K-feldspar.

41) Fig. 5.18; not VM but from same rock as VM GeoLab M05; Calcite with exsolved dolomite; from Glenelg, Inverness-shire; NW Scotland; 1000 Myr; contains calcite, dolomite.

Part 2: Alphabetic list of minerals. Each mineral name is followed by the number or numbers for the thin sections (from Part 1) in which that mineral is present.

Albite and Na-rich plagioclase: 1, 9, 10, 14, 16, 29, 35, 37, 39,

Andalusite: 7, 28,

Anthophylite: 22,

Apatite: 1

Augite (variety of clinopyroxene): 12, 23,

Biotite: 1, 4, 5, 7, 9, 10, 27, 28, 35, 38,

Brucite: 26,

Calcite: 8, 26, 41,

Chlorite: 3, 9, 19, 29, 31,

Chrome spinel: 18, 20,

Clinopyroxene in general: 18, 24, 25, 37,

Coesite: 30

Cordierite: 7,

Corundum: 27

Diopside (variety of clinopyroxene): 8,

Dolomite: 8, 21, 41

Epidote: 9, 10, 13,

Garnet: 1, 2, 3, 4, 5, 11, 13, 14, 29, 30, 35, 37,

Glaucophane: 13, 29,

Hornblende: 10, 11, 14, 15, 37, 39,

K-feldspar: 2, 17, 35, 40,

Kyanite: 5, 35, 38, 40,

Magnesite: 21, 22,

Muscovite: 1, 3, 4, 5, 28, 29, 38,

Olivine (including forsterite): 8, 18, 19, 34,

Omphacite (variety of clinopyroxene): 14, 39,

Opaques: 1, 15, 20, 21, 22,

Orthopyroxene (including enstatite): 12, 18, 19, 24,

Perthitic feldspar: 2, 36,

Plagioclase (intermediate and Ca-rich) 11, 12, 15, 18, 23, 24, 32, 33,

Quartz: 1, 2, 3, 4, 5, 7, 9, 10, 14, 16, 17, 28, 30, 35, 37, 38, 40,

Ringwoodite: 34

Rutile: 14, 37,

Serpentine: 8, 20,

Sillimanite: 6, 38,

Staurolite: 4,

Talc: 19, 21,

Titanite: 10,

Tourmaline: 4

Appendix 4 Microbeam and X-ray methods

A4.1 The scanning electron microscope (SEM)

A4.1.1 Kinds of image produced by the SEM

The **scanning electron microscope** (abbreviated SEM) is a versatile and powerful analytical tool for examining natural or polished rock surfaces over a range of magnifications from about 10 times to 500,000 times. It provides three separate kinds of image known as secondary electron images, back-scattered electron images, and X-ray element maps. The last two complement the images obtained with a polarizing microscope.

1 A secondary electron image shows what one might expect to see through a very powerful magnifying glass. It shows the three-dimensional appearance of a sample such as the broken surface of a rock specimen (e.g. the slate in Fig. 2.3).

2 A back-scattered electron image can also show a surface in three dimensions, but its main use is to show the different minerals in a polished thin section. Each mineral appears in a different shade of grey. Paler shades correspond to 'heavier' minerals, such as those containing a good deal of iron (Fig. A4.1, panel B).

3 An X-ray element image is also obtained from a polished thin-section surface. Here the brightness corresponds to the amount of a chosen element, for example aluminium, that is present in each mineral, and the image can be in colour rather than grey. A set of X-ray images of a chosen area can be obtained, one for each element (e.g. Fig. A4.1 panels C, D and E). Each image is an element distribution map for a tiny area of the polished rock surface. Two or more coloured X-ray maps, when superimposed, will generally show the different minerals, each in a combined colour (Fig. A4.1, panel F).

A4.1.2 How does an SEM work?

An SEM works quite differently from a light microscope. A beam of high-energy electrons (typically accelerated by 15 or 20 kilovolts) is focused by magnetic lenses onto a really tiny spot where it hits the surface of the specimen. The position of the spot can be deflected backwards and forwards, and from left to right, by adjusting the voltages on two pairs of electrostatic plates between which the beam passes (Fig. A4.2). When taking a picture, the voltages on the plates are controlled by a computer to make the beam scan rapidly and repeatedly in a **raster pattern** over a chosen rectangular area on the surface. A raster pattern is where the tip of the beam systematically traces a series of left-to-right lines, each line off-set to lie beside the previous line. The scanned rectangle is somewhat analogous to the straight swathes of grass on a freshly mown lawn.

The three different kinds of image are produced with the aid of three detectors, positioned close to the specimen. One detector monitors secondary electrons, another registers back-scattered electrons, and the third picks up X-rays (Fig. A4.2).

Secondary electrons are electrons dislodged from the surface as the primary electron beam passes over it. They are launched at a low speed, i.e. with low energy. The detector is located to one side of the incoming electron beam. As the beam moves over the uneven surface, the detector will pick up many more secondary electrons from a part of the surface that slopes towards it, as in position (A) in Figure A4.3, than from a part that slopes away, as in position (B). The signal from the detector is amplified and fed to a computer; the signal controls the brightness of a spot that is rastering over the whole area of the computer's display screen, exactly synchronized with the movement of the rastering electron beam over the chosen tiny rectangle on the specimen. The result is that brighter and darker regions on the display screen replicate perfectly the detailed three-dimensional appearance of the specimen at a high magnification, as though the specimen were illuminated by a light source from the side where the detector sits. The magnification is the ratio of the width of the computer's display screen to the width of the scanned rectangle on the specimen, so the magnification can be increased simply by scanning over a smaller area of the specimen.

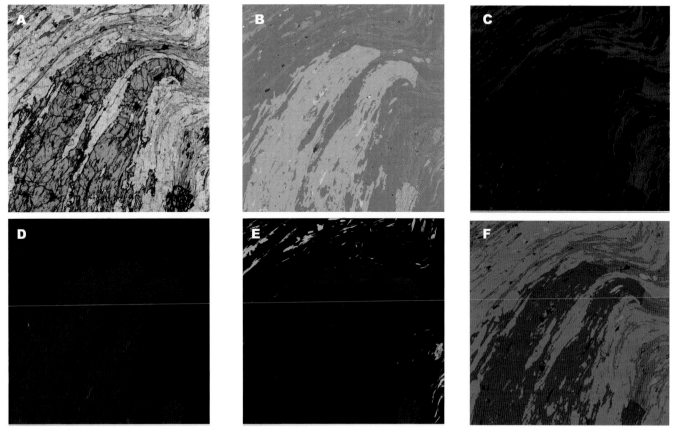

Figure A4.1 Montage of six images of the same tiny area of a polished thin section of the garnet-mica schist, M08, described in chapter 2 (Fig. 2.21). (**A**) is in PPL using a polarizing microscope, and shows garnet replicating the shape of an earlier fold by replacing a muscovite-rich layer in the foliation. (**B**) is a backscattered electron (BSE) image. Garnet is pale grey, whereas quartz and muscovite are both dark grey and are hard to tell apart. Small, medium grey elongate grains near the top are chlorite. Very pale grey inclusions in the garnet are iron and titanium oxides. (**C**) is an X-ray element map for aluminium. Bright orange is muscovite. Garnet is dull orange. Quartz (with no Al) is black. (**D**) is an X-ray map for iron, which is almost entirely in garnet. (**E**) is the same for magnesium, which is mostly in the small plates of chlorite at the top, but can also be seen faintly in the garnet, where its concentration increases slightly towards the upper end of the elongate, hook-shaped grain. (**F**) is a composite X-ray element map. Garnet is purple, quartz dirty yellow, muscovite orange, and chlorite green. Deep orange patches within the muscovite are Na-plagioclase. Bright blue inclusions in the garnet are the accessory mineral apatite (calcium phosphate).

Back-scattered electrons are those that penetrate a short distance into the specimen and are then strongly deflected as they pass close to the central heavy nuclei of atoms. They loop around without slowing down and come flying back out of the specimen. The detector is tuned to these 'high-energy' electrons. It sits flat, immediately above the specimen, and is disc-shaped with a central hole to allow the incoming electron beam to pass through. The proportion of the electrons that are back-scattered depends on the average atomic weight of the elements in the mineral being targeted, so the rastering spot on the computer's display will turn brighter when the electron beam is passing over a 'heavy' mineral, and turn darker when it is over a less dense mineral, as shown in the garnet-mica schist in Figure A4.1.

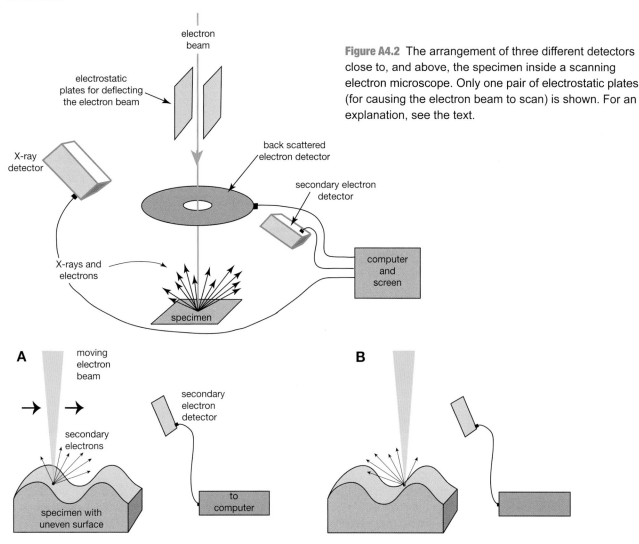

Figure A4.2 The arrangement of three different detectors close to, and above, the specimen inside a scanning electron microscope. Only one pair of electrostatic plates (for causing the electron beam to scan) is shown. For an explanation, see the text.

Figure A4.3 Diagram showing how the number of secondary electrons reaching the detector depends on the slope of the specimen surface. In **A**, where the surface slopes to the right, the detector receives more secondary electrons than in **B**, where it slopes to the left.

X-rays are generated when the high-speed electrons in the beam interact with electrons close to the nuclei of the atoms through which they pass. Each chemical element produces its own so-called characteristic X-rays that have distinctive, fixed values of energy. The X-ray detector is able to pick out these characteristic X-rays from the background X-rays and they appear, after amplification and processing in the SEM's computer, as separate sharp peaks on an X-ray energy spectrum. Each element has a main peak and often one or more minor peaks. To produce an X-ray element map, the main peak for a chosen element, e.g. the main X-ray peak for aluminium, is isolated from the X-ray spectrum and monitored separately. As the electron beam scans over the polished surface of the specimen, the height of this peak changes from mineral to mineral, and controls the brightness of the moving spot on the computer's screen (Fig. A3.1, panel C). So in this case, minerals with abundant aluminium will appear bright, and those with little aluminium will appear dim.

Examples of energy spectra obtained from three different minerals in the garnet-mica schist, M08, are shown in Figure A4.4. The element symbols for the main X-ray peaks are labelled. The peaks are in the same order as the elements in the Periodic Table (Fig. A2.1 in Appendix 2), i.e. O, Na, Mg, Al, Si, K, Ca and Fe.

Quartz has peaks for O and Si, though note that the O peak is *not* twice the size of the Si peak, as the formula SiO_2 might lead one to expect. It is actually a lot smaller – about one-sixth of what one might expect. Clearly, element peak heights do not correspond directly with a mineral's formula. Chlorite has peaks for O, Mg, Al, Si and Fe, consistent with, but again not in proportion to, its mineral formula (Table A2.1) and H, which is present as (OH) in chlorite, is not detected at all. Garnet contains abundant Fe but it also contains Ca, a little Mg, and a small but significant amount of the minor element manganese (Mn), in line with the limited atomic substitution in this mineral (see section A2.4.1, and Table A2.1).

A4.1.3 Electron probe micro-analysis (EPMA)

As well as producing X-ray maps, the SEM can produce accurate chemical analyses of chosen parts of individual mineral grains. Used in this mode, the electron beam scans a tiny rectangle on a single grain, or remains stationary, focused to a sharp point. The X-rays from the chosen area or spot are collected and accumulated over a period of perhaps one minute, and the resulting energy spectrum is converted by the computer into the chemical composition of the mineral at that position. When used like this, the SEM is said to be operating as an **electron probe micro-analyser (EPMA)**. EPMA compositions of minerals are widely reported in studies of metamorphic rocks, and examples of their use in inferring metamorphic temperatures are given in chapter 5. The conversion of an X-ray spectrum to a mineral formula is a sophisticated, and non-trivial, calculation based on the physics of X-ray generation and X-ray transmission.

A4.2 X-ray powder diffraction (XRD)

X-ray powder diffraction is a destructive technique for identifying minerals. It is not for determining their chemistry. The equipment used is called a powder diffractometer, and a

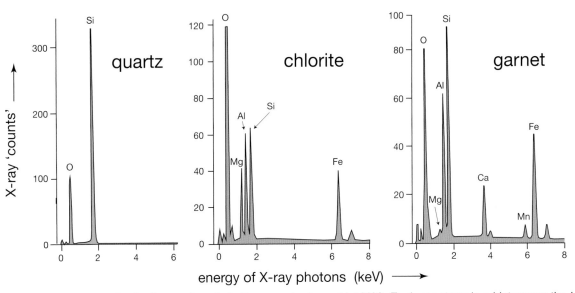

Figure A4.4 X-ray energy spectra for three minerals in the garnet-mica schist M08. Each spectrum is a histogram: the horizontal axis is X-ray energy measured in units of a thousand electron volts (keV), and the vertical axis is the number of X-ray 'particles' (called photons) detected and 'counted' per second for the peak.

134

small sample of pulverized rock or mineral is placed into its specimen holder. The arrangement of its parts is shown in Figure A4.5.

X-rays from an X-ray generator are passed through a narrow slit to produce a thin beam that impinges on the powdered sample. Some X-rays are scattered in new directions by the sample, and picked up by an X-ray detector attached to a pivoting arm called a goniometer. The detector records the varying 'brightness' of the scattered X-rays as the goniometer turns and the angular position of the detector slowly changes. The scattering is not random, but happens only at very specific angles inclined to the direction of the incoming X-ray beam (Fig. A4.5). The angles are called 2-theta angles (written 2Θ). The detector's output appears as a series of narrow peaks on a chart. An example of a recording for a sample of powdered slate is shown in Figure A4.6, and is discussed in chapter 2, section 2.2. Each mineral has its own distinctive *set* of peak positions (2Θ angles) and brightness values, which together provide a 'fingerprint' for that mineral and allow it to be identified with certainty. The slate sample in Figure A4.6 has sets of peaks for the three minerals – quartz, chlorite and muscovite – which are labelled.

To add an incidental historical note here, the physics behind X-ray scattering by crystals was explained in a very elegant way in 1913 by W.H. and W.L. Bragg (father and son). In 1915 they jointly won the Nobel Prize for Physics for their work, which revealed the crystal structures of many minerals, including the silicates whose structures were discussed in Appendix 2. Their work led later to the discovery in 1953 of the

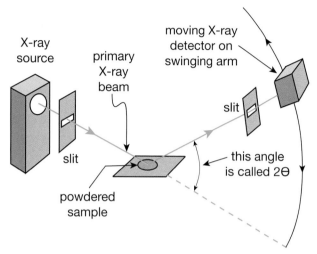

Figure A4.5 Schematic layout of the parts of an X-ray powder diffractometer.

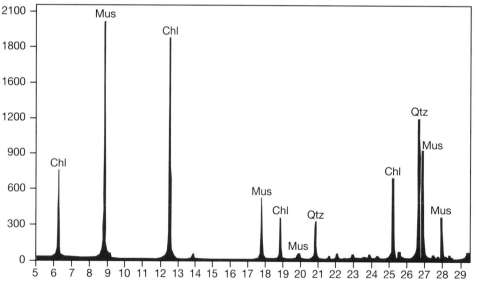

2Θ (angle in degrees of X-ray detector relative to primary X-ray beam)

Figure A4.6 X-ray powder diffraction chart for a piece of slate containing quartz (Qtz), muscovite (Mus) and chlorite (Chl). The horizontal axis shows the angle (called 2-theta) through which the X-ray beam is deflected when it is scattered by the mineral grains in the powdered sample. The vertical axis shows the 'brightness' (counts) of the X-rays at each angle. Each mineral has its own distinctive set of angles and brightness values.

double helix structure of DNA in the Cavendish Laboratory, Cambridge, where W.L. Bragg was director. He is captured on film (https://www.youtube.com/watch?v=UEB39-jlmdw) giving the 1952 Royal Institution Christmas Lecture on crystal behaviour.

What the Braggs achieved in 1913 was to show how regularly spaced layers of atoms within a piece of crystalline material act like tiny mirrors that reflect the incoming X-ray beam. They showed that the mirrors only 'work' if the angle of reflection, Θ, is just right. The angle depends on the wavelength, λ, of the X-rays and the distance, d, separating adjacent layers of atoms, according to the Bragg equation: $n\lambda = 2d \sin\Theta$ ('n' is a whole number).

Appendix 5

The principles of isotopic dating (geochronology)

This appendix briefly outlines the principles of **isotopic dating**. It focuses on the two dating methods singled out for discussion in chapter 4 (section 4.2.3), namely the uranium–lead dating of zircon crystals, and the potassium–argon dating of biotite crystals.

A5.1 Uranium–lead dating of zircon crystals

Zircon is an accessory mineral containing the rare element zirconium (Zr). Its formula is $ZrSiO_4$. Tiny prisms of it, too small to see without a microscope, grow from magma as it crystallizes, and individual zircon prisms can be separated from the resulting igneous rock (Fig. A5.1), and dated.

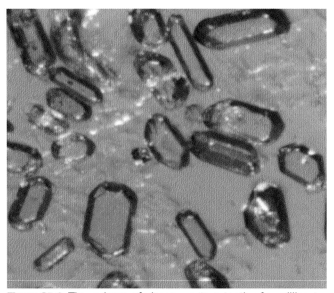

Figure A5.1 Tiny prisms of zircon, up to a tenth of a millimetre long, separated from a sample of granite in the west of Ireland. Sieved grains of crushed granite were immersed in a very dense liquid called methyl iodide. Quartz, feldspar and mica float in this liquid, while the zircon grains, being extremely dense, sink and can be separated. *Photo courtesy of Stephen Daly.*

As zircon crystals grow they take in atoms of uranium (U) from the magma. Uranium is a heavy element that is even rarer than zirconium. It substitutes readily for zirconium within the zircon crystals. It amounts to perhaps 0.05% of the zircon by weight. Now, uranium is **radioactive** and decays (changes) very, very slowly to lead (contributing, incidentally, to the Earth's heat budget as it does so). Therefore, ancient zircon crystals now contain lead (Pb) that has come from the decay of uranium. In fact, as zircon crystals grow they virtually exclude lead because Pb atoms are the wrong size to substitute for Zr atoms. Thus, the lead in an ancient zircon crystal today will have come almost exclusively from uranium.

The rate at which uranium decays to lead has been measured accurately using pure uranium compounds and an instrument like a Geiger counter, which can detect the tiny pulse of energy released as each U atom decays. Also the minuscule amount of uranium and the even smaller amount of lead in a zircon crystal can be measured precisely with the help of an instrument called a mass spectrometer. So, by knowing the amount of uranium still present in a zircon crystal, and the amount of lead that has built up there, and also knowing the rate at which uranium changes to lead, it is possible to calculate the time that has elapsed since there was no lead in the zircon crystal. This is the crystal's age.

The concept of the 'rate of decay' of uranium needs a little clarification. The uranium does not decay at a steady, uniform rate so that after a period of time it will all have gone; rather it decays in a so-called exponential way, halving in amount after the passage of a fixed period of time known as the **half-life** of uranium. The half-life of uranium is about 4500 million years (about the same as the age of the Earth, incidentally), and the mathematical formula used to calculate the age of a zircon crystal, in millions of years, is:

Age $= 14840 \times \log_{10}(1 + $ ratio of lead/uranium)

The ratio used in this formula is that of the *numbers* of lead and uranium atoms, not the ratio of their weights.

In the above discussion, uranium and lead ought strictly to have been referred to as uranium-238 and lead-206 (written as ^{238}U and ^{206}Pb, respectively). These are so-called **isotopes** of uranium and lead. It is these isotopes that are measured using a mass spectrometer. As it happens, 99.3% of all uranium is ^{238}U; the last little bit, 0.7%, is another isotope called uranium-235 (^{235}U). So if one were to measure the total uranium and the total lead in a zircon crystal, instead of measuring the amounts of ^{238}U and ^{206}Pb, one would still get a pretty good estimate of the age. In fact, the first serious attempts to date rocks were done in this way over 100 years ago (only a few years after radioactivity had been discovered) by a remarkable geologist and physicist named Arthur Holmes. In 1911 he showed, from his measurement of lead in a uranium ore in a Devonian rock from Norway, that the ore was 370 million years old. This age is very close to what is now known to be the correct age of the ore, and it was obtained ten years before isotopes had even been discovered, and decades before mass spectrometers were invented.

By good fortune, zircon crystals can also be dated using the other isotope of uranium, ^{235}U, which, again, is radioactive. It decays at a known rate to lead-207 (^{207}Pb). Although ^{235}U is far less abundant than ^{238}U, its tiny quantity in a zircon crystal is measurable, and so is the infinitesimal amount of ^{207}Pb to which it changes. So it is possible to obtain the age of a zircon crystal in two independent ways – from the ratio of ^{207}Pb to ^{235}U and from the ratio of ^{206}Pb to ^{238}U. If both ways give the same age, which they usually do, the age is called a concordant uranium-lead age, and this confirms that the age is accurate.

Uranium–lead dating today has become quite sophisticated. Technology has now advanced to the point where it is possible to date separate *parts* of an individual zircon crystal in a polished thin section using a special mass spectrometer called a laser ablation inductively-coupled plasma mass spectrometer, abbreviated LA-ICPMS. The polished thin section is placed inside the instrument, and the chosen area of zircon to be dated is targeted with a powerful laser beam, causing ablation (evaporation) of the crystal leaving a shallow pit in the zircon's surface. The atoms released to the vapour are analysed by the mass spectrometer. This method of U-Pb dating can also be applied to uranium-bearing minerals other than zircon, and it has significantly increased the potential for constraining little 't' in a metamorphic P–T–t path.

A5.2 Potassium–argon dating of biotite

The second dating technique selected here to illustrate principles is the K–Ar method for dating biotite. The basis for this method is that the isotope of potassium called potassium-40 (^{40}K) decays at a known rate to produce the daughter isotope, argon-40 (^{40}Ar). So, by measuring the amounts of the parent (^{40}K) and daughter (^{40}Ar) isotopes in a sample of biotite, one can calculate the time elapsed since the daughter isotope, ^{40}Ar, started to accumulate in the biotite.

Potassium-40 is scarce. It amounts to only 0.0117% of all potassium atoms, but this percentage is fixed, so the amount of ^{40}K can be calculated simply from the total measured amount of potassium in the biotite sample. Argon-40 is a gas; it is extracted by heating the biotite, then the amount of it is measured with the help of a mass spectrometer. This method does not date the *formation* of the biotite; rather it records the time in the biotite's cooling history when the temperature dropped below about 300°C. Three hundred degrees centigrade is the so-called **blocking temperature** for argon in biotite.

The concept of a blocking temperature is illustrated in Figure A5.2. Argon is an inert gas, but its atoms are large, so, if the temperature is not too high, when a new ^{40}Ar atom is formed it will generally remain lodged in the place formerly occupied by its ^{40}K parent atom. It cannot easily fit through the gaps between the surrounding atoms. However, at a temperature hotter than about 300°C biotite is leaky to argon, so the ^{40}Ar escapes as fast as it is produced. This is the case at the peak of metamorphism when the temperature for the formation of biotite schist (in the biotite zone or higher) is more than about 400°C. Later, following the orogeny, at the point where the biotite cools below 300°C, the argon stops leaking and begins to accumulate; this is the stage when the K–Ar clock starts ticking.

The K–Ar chronometer is often used in a versatile and clever way that involves first converting some of the stable potassium isotope, ^{39}K, to ^{39}Ar by irradiating the mineral in a nuclear reactor. Argon is then extracted by heating the mineral and the ratio of ^{40}Ar to ^{39}Ar is measured. Since the ^{39}Ar is proportional to the amount of potassium (and hence ^{40}K) in the mineral, the ratio $^{40}Ar/^{39}Ar$ can be converted to the mineral's K–Ar age. This method is called $^{40}Ar/^{39}Ar$ dating.

$^{40}Ar/^{39}Ar$ dating has become remarkably sophisticated. By combining it with the laser ablation technique (LA-ICPMS),

138

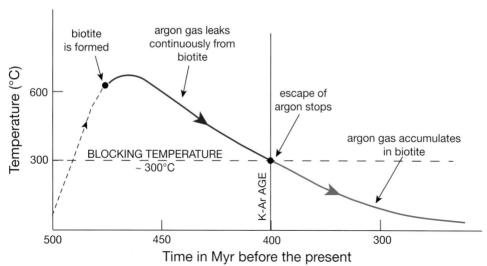

Figure A5.2 Diagram showing how a K–Ar age from biotite is interpreted as a 'cooling age', recording the time elapsed since the mineral cooled below about 300°C. For an explanation see the text.

mentioned above in the context of U–Pb dating, it can be used to obtain $^{40}Ar/^{39}Ar$ ages of chosen parts of individual mineral grains in a polished thin section that has been appropriately irradiated. Moreover, some minerals actually grow at a temperature *below* their blocking temperature, particularly during low-grade metamorphism. Radiogenic ^{40}Ar will have accumulated inside these minerals from the outset, and the date of mineral *growth*, rather than the date of cooling through the blocking temperature, will be obtained. Constraining little 't' of a rock's P–T–t path by methods such as this is very much at the forefront of investigating metamorphic rocks from orogenic belts today.

Glossary

A

absorption colour [122]: the formal name for the colour of a mineral in thin section seen in PPL.

accessory mineral [2, 117, 127]: a mineral that is present in a small amount in a rock, and is not an essential constituent of the rock. It usually contains a high content of a rare element.

ACF triangle [36]: an equilateral triangle for showing graphically the compositions of rocks or minerals in terms of the three elemental constituents, aluminium, calcium and combined iron and magnesium.

acidic (igneous rock) [41]: an igneous rock with more than 63% by weight of SiO_2. Granite and rhyolite are acidic igneous rocks. The term felsic is commonly used instead of acidic.

actinolite [31, 114]: clinoamphibole with the formula $Ca_2(Mg,Fe)_5Si_8O_{22}(OH)_2$.

albite [30, 111, 115]: sodium feldspar, $NaAlSi_3O_8$. An end-member of the plagioclase solid solution series.

almandine [18, 113]: iron garnet, $Fe_3Al_2(SiO_4)_3$.

amphiboles [114]: a group of silicate minerals with double chains of linked SiO_4 tetrahedra running parallel to each other, and having the repeat unit $Si_4O_{11}(OH)^{7-}$, which is usually doubled and written $Si_8O_{22}(OH)_2^{14-}$. Atoms of all other main elements of the continental crust can supply the positive charges. Amphiboles form prismatic crystals, and have two cleavage directions intersecting at about 60°.

amphibolite [33]: a common kind of metabasite that is dark green to black and composed of the minerals hornblende and Ca-bearing plagioclase.

anatexis [22]: see partial melting.

anchimetamorphism [5]: very low-grade metamorphism corresponding to diagenesis between about 200°C and 300°C. It is defined by the extent to which new muscovite has crystallized in shale, as determined by X-ray diffraction.

anchizone [5]: a part of a metamorphic belt affected by anchimetamorphism.

andalusite [23, 113]: a pale-coloured silicate with prismatic crystals and formula Al_2SiO_5. It is one of three polymorphs of Al_2SiO_5, the others being kyanite and sillimanite.

andesite [41]: a fine-grained intermediate igneous rock.

anhedral [62]: the shape of a mineral grain that has no crystal faces, in contrast to a euhedral grain shape.

anorthite [29, 111, 115]: pure calcium feldspar, $CaAl_2Si_2O_8$. An end-member of the plagioclase solid solution series.

anorthosite [86]: a coarse-grained igneous rock made from plagioclase. It is the rock from which the lunar crust is made.

anthophyllite [48, 114]: the magnesium end-member of orthoamphibole, formula $Mg_7Si_8O_{22}(OH)_2$.

apatite [117]: calcium phosphate, a widespread accessory mineral.

aragonite [26]: a polymorph of $CaCO_3$, less common and more dense than calcite.

asbestos [47]: any silicate mineral that breaks into fibres when rubbed or crushed.

asthenosphere [105]: the mantle below the lithosphere.

atomic substitution [17, 110]: a feature of the chemical composition of some minerals in which an atom (strictly *ion*) of one element, e.g. Mg^{2+}, can be substituted by a similar sized atom (ion) of another element, in this case Fe^{2+}.

augen [18]: (plural noun) large lens-shaped mineral grains visible in some foliated metamorphic rocks. In cross-section they look like eyes. Common minerals that form augen are garnet and feldspar.

augen gneiss [42]: gneiss with abundant augen. It is usually granitic with augen of feldspar.

augite [30, 113]: clinopyroxene with the formula $Ca(Mg,Fe)Si_2O_6$ modified by limited atomic substitution of Na for Ca, Mg for Ca, Al for Mg, and Al for Si.

B

Barrow zone [21]: see metamorphic zone.

basalt [1, 104]: a common, dark-coloured, fine-grained basic igneous rock. It is composed chiefly of the minerals augite and Ca-rich plagioclase feldspar. It forms when basic magma cools rapidly, usually by erupting onto the Earth's surface or the seabed.

basic (igneous rock) [30]: a term used to describe an igneous rock whose silica content falls between 45% and 52% by weight. Basalt, dolerite and gabbro are basic igneous rocks.

batholith [74]: a cluster of contiguous plutons.

bed (bedding) [15]: an original layer (original layering) of sedimentary rock.

Becke test [115, 126]: a test using the polarizing microscope to help distinguish between different minerals with low relief.

biotite [16]: a black sheet silicate with perfect cleavage and the formula $K(Mg,Fe)_3Si_3AlO_{10}(OH)_2$. Along with muscovite it is a variety of mica.

birefringence [121]: an optical property related to refractive index that helps to determine interference colours of mineral grains in XP.

black smoker [8]: a rising plume of superheated dark cloudy water streaming from a vent on the seabed into the cold seawater. Such plumes occur on spreading ridges.

blastomylonite [43]: mylonite that has later recrystallized.

blocking temperature [80, 137]: the temperature below which a radiogenic element becomes trapped, or blocked, within its host mineral.

blueschist [34]: a kind of metabasite that is composed of the minerals amphibole (the Na-Al-rich variety called glaucophane) and a hydrous mineral containing calcium and aluminium, either epidote or lawsonite.

C

calcite [24]: the common mineral form of calcium carbonate, $CaCO_3$.

calc-silicate rock [29]: a metasediment derived from marl or siliceous dolomite containing little or no carbonate.

carbonate ion [112]: a tightly bonded group of three oxygen atoms and one carbon atom. It carries two negative charges and has the formula $(CO_3)^{2-}$.

cataclasis [9]: the crushing and grinding of pre-existing rocks during dynamic or shock metamorphism.

cataclasite [9]: a rock produced by cataclasis.

chlorite [15, 115]: a green, water-rich sheet silicate with substantial substitution of Fe for Mg, Al for Mg and Al for Si, giving the formula $(Mg,Fe,Al)_3(Si,Al)_4O_{10}(OH)_2.3(Mg,Fe,Al)(OH)_2$.

chondrite [44, 104]: the most common kind of meteorite, made largely of grains of peridotite, and thought to be representative of the material from which the Earth as a whole was made.

chrome spinel [44]: a dense, brown accessory oxide mineral, $Mg(Al,Cr)_2O_4$.

clay minerals [14]: a group of sheet silicates that occur as tiny particles in clay and shale.

cleavage (in minerals) [15, 113, 124]: the potential for a mineral to break along parallel planes, called cleavage planes. Depending on the number of directions of cleavage planes it has, a mineral may break into sheet-like, elongated, or blocky pieces, with flat surfaces.

clinoamphibole [32, 114]: amphibole with 8-fold and 6-fold (octahedral) sites between the chains in the ratio 2:5. A simple formula is $Ca_2(Mg,Fe)_5Si_8O_{22}(OH)_2$.

clinopyroxene [30, 113]: pyroxene with equal numbers of 8-fold and 6-fold (octahedral) sites between the chains. A simple formula is $Ca(Mg,Fe)Si_2O_6$. Specific examples are diopside, augite and omphacite.

close-packed structure [113]: an arrangement of atoms in planes in a crystal structure, exemplified by the oxygen atoms in olivine. Each atom touches six others in the same plane, and touches three atoms in the plane below and three in the plane above.

coesite [82]: a high-pressure polymorph of SiO_2.

conduction (of heat) [77]: the flow of heat through a body of material in response to a temperature difference between one place and another. Heat flows from a hot place to a cooler place.

contact aureole [7]: the border of country rock surrounding an igneous intrusion where contact metamorphism has clearly taken place.

contact metamorphism (= thermal metamorphism) [7]: metamorphism caused by the close proximity of a hot body of igneous rock.

continental crust [103, 109]: the outer layer of the solid Earth beneath continents and neighbouring shallow seas. It is, on average, 35km thick and its base is defined by the Moho.

continuous (dehydration reaction) [55]: a metamorphic reaction yielding water continuously over a range of increasing temperatures.

convection [8, 59]: the circulation of fluid driven by density contrast. Where part of a body of fluid is heated, its density decreases and it rises, while a neighbouring part is cooler and denser and so sinks to replace the rising fluid.

coordination number (with oxygen) [112]: the number of oxygen atoms that touch a positive ion, making a polyhedron. It is written 4-fold, 6-fold, 8-fold, etc. depending on the coordination number.

cordierite [23 115]: a ferromagnesian framework silicate with the formula $(Mg,Fe)_2(Si_5Al_4)O_{18}$.

corundum [72]: the mineral name for aluminium oxide, Al_2O_3. The blue variety is called sapphire.

country rock [7]: the rock hosting an igneous intrusion.

coupled atomic substitution [99, 111]: atomic substitution where exchanging atoms (as ions) have different charges, so that two separate substitution systems, such as Ca^{2+} for Na^+ and Al^{3+} for Si^{4+}, must operate simultaneously in order to retain a neutral charge in an intermediate mineral formula between two end-members.

crenulations [17]: miniature corrugations of the schistosity in schist or phyllite.

crossed polars [120]: see XP.

crust [1]: see continental crust and oceanic crust.

crystal [2]: a crystalline solid bounded by flat surfaces, called crystal faces, whose orientations are determined by the regular (crystalline) arrangement of the atoms throughout its interior.

crystal structure [112]: see crystalline.

crystalline [12]: an adjective describing solids with a regularly repeating internal arrangement of atoms called the crystal structure. All minerals are crystalline.

crystallization (of magma) [44, 80]: change on cooling from a liquid to a crystalline state. Crystallization is the same as solidification retain a neutral charge in an intermediate mineral formula between two end-members.

cumulates [44]: rocks formed by settling out of crystals from magma.

D

decompression melting [107]: the melting or partial melting of rock in response to a reduction in pressure.

deformation twin lamellae [27]: sets of narrow parallel bands seen under the microscope in calcite and dolomite.

diagenesis [5, 26]: the process of change in sedimentary rocks in the subsurface, generally at temperatures below about 300°C. It includes the process of lithification and is gradational with increasing temperature into metamorphism.

diaplectic glass [86]: glass produced from the shock melting of individual minerals.

diopside [26, 113]: the magnesium end-member of clinopyroxene, $CaMgSi_2O_6$.

diorite [41]: a coarse-grained intermediate igneous rock.

directed stress [4]: the maximum compressive force in solid rocks beneath the surface in places where the rocks are being squeezed more in one direction than in other directions. This causes a body of rock to change in shape, i.e. to undergo strain.

discontinuous (dehydration reaction) [55]: a metamorphic reaction releasing water in one go as temperature rises.

dolerite [30, 104]: a medium-grained basic igneous rock, coarser than basalt.

dolomite [26, 109]: a carbonate mineral with equal calcium and magnesium, $CaMg(CO_3)_2$.

double chain silicate [114]: see amphibole.

dunite [44]: peridotite made entirely of olivine.

dyke [30]: a sheet-like intrusion of igneous rock that is not parallel to layering in the country rock.

dynamic metamorphism [9, 41]: a localized variety of metamorphism that occurs in fault zones or shear zones. It results in the crushing and grinding of mineral grains.

E

eclogite [35, 97, 113]: a dense metabasite composed of red garnet and green Na-Al rich clinopyroxene named omphacite.

electron probe micro-analysis [94, 133]: a technique involving an electron beam and X-rays for measuring the chemical composition of a part of a mineral grain in a polished thin section.

end-member [18, 110]: see solid solution.

enstatite [44, 113]: the magnesium end-member of orthopyroxene, $MgSiO_3$.

epidote [29, 113]: a greenish-yellow, dense silicate mineral, formula $Ca_2(Al,Fe^{3+})_3(SiO_4)_3(OH)$.

epidote amphibolite [32]: a kind of metabasite similar to amphibolite, but containing sodium-rich plagioclase (albite) and epidote instead of Ca-bearing plagioclase.

equilibrium (= stability) [52]: a state of minimum energy. A metamorphic rock is in equilibrium, or is stable, at a particular pressure and temperature when it has minimized its excess energy (strictly, its excess Gibbs energy) which includes the chemical energy of its minerals and the surface energy of its grains.

euhedral [18, 62]: displaying good crystal faces.

eutectic [59]: the first melt to form when partial melting begins. The term refers both to the composition of the melt, and to its temperature.

experimental petrology [39, 54]: the synthesis of rocks and minerals under controlled high pressure and temperature conditions in the laboratory.

exsolution [91]: the nucleation and growth of tiny grains of one mineral within another, as in the formation of perthite. This happens when falling temperature leaves a mineral composition outside the limit of permitted atomic substitution.

extensional sedimentary basin [108]: a huge depression on the Earth's surface where sediment accumulates. It is caused by stretching and thinning (extension) of the lithosphere, including the continental crust within it.

extinction (in a polarizing microscope) [20, 122]: a state of blackness seen in mineral grains in XP, that occurs four times during one rotation of the stage.

F

facies [5]: see metamorphic facies.

fault breccia [9]: a jumbled mixture of angular rock fragments and rock powder produced by fault movement not far beneath the surface.

fault gouge [9]: rock powder associated with fault breccia.

fayalite [110]: pure iron olivine, Fe_2SiO_4. An end-member of the olivine solid solution series.

feldspar [115]: an important group of framework silicates that fall into two subgroups, plagioclase feldspar and potassium feldspar.

Fe-Mg exchange thermometer [93]: a geothermometer based on the way in which Fe and Mg are distributed between two minerals in a metamorphic rock, with one mineral having higher Fe/Mg than the other.

fluid (= fluid phase) [2]: a liquid or a gas which, in the context of metamorphism, is usually dominated by H_2O with dissolved salts. If present, it permeates along the boundaries between grains and

facilitates grain growth. It may contain, or even consist entirely, of CO_2. It is commonly consumed, or released, by chemical reactions between minerals as they cool down or heat up, respectively, during metamorphism.

fluid inclusion [2]: a tiny cavity entirely enclosed by a mineral grain and filled with fluid, and in some cases bearing crystals precipitated from the fluid.

foliation (= foliated texture) [3]: a loosely defined term for a rock texture where grains with flattened shapes are aligned or where there is banding. It implies metamorphism under the influence of directed stress, and includes the textures known as slaty cleavage, schistosity and gneissic banding.

forsterite [26, 110]: pure magnesium olivine, Mg_2SiO_4. An end-member of the olivine solid solution series.

framework silicates [115]: a group of silicate minerals in which SiO_4 and AlO_4 tetrahedra are linked at all their corners to give a three-dimensional structure described as a framework. Quartz, plagioclase, and K-feldspar are examples.

G

gabbro [30, 104]: a coarse-grained basic igneous rock identical in composition to basalt.

garnet [17, 113]: a hard, dense, glassy-looking red mineral often with a shape known as a rhombic dodecahedron. Its formula is $(Fe,Mg,Ca)_3Al_2(SiO_4)_3$.

geobarometry [89]: the science of estimating the pressure at which a metamorphic rock was formed.

geochronology [79]: the science of determining dates, usually in millions of years (Myr) before the present, for past geological events.

geotherm (= geothermal gradient) [107]: the natural increase of temperature with depth within the Earth. In the upper continental crust it is commonly about 20°C per km.

geothermobarometry [89]: geothermometry and geobarometry combined.

geothermometry [89]: the science of estimating the temperature at which a metamorphic rock was formed.

Gibbs energy [52]: see equilibrium.

glaucophane [34, 114]: blue-to-lilac clinoamphibole in which substantial Na substitutes for Ca, and the same amount of Al substitutes for Mg, giving the formula $(Na,Ca)_2(Mg,Fe,Al)_5Si_8O_{22}(OH)_2$.

gneiss [3, 22]: a poorly defined name for a high-grade metamorphic rock that is coarse-grained, usually banded or foliated in appearance and usually containing feldspar. It can be derived from various protoliths, including igneous rocks, sandstone and shale.

grade [5]: a term referring to the temperature (but not the pressure) at which metamorphism occurs. Rocks are described as being low-grade, medium-grade and high-grade when formed, respectively, in the approximate temperature ranges 300–500°C, 500–700°C, and 700–900°C.

grain growth (= recrystallization) [3]: a metamorphic process whereby mineral grains increase in size. It implies that many small grains of a mineral coalesce to make a few larger grains.

granite [1, 41]: It includes the process of lithification and is gradational with increasing temperature into metamorphism. It usually forms when magma, rising from a source at depth in the continental crust, fails to reach the surface and inflates an underground space for itself, called a magma chamber, where it cools slowly and

141

solidifies as a pluton.

granoblastic [25, 34]**:** a term describing a metamorphic texture in which grains are all about the same size and have polyhedral shapes (or polygonal outlines in a thin section). Such a texture indicates a high level of equilibration.

greenschist [31]**:** a common kind of foliated metabasite that is dark green and composed mainly of the minerals chlorite, epidote, and Na-rich plagioclase (albite).

greenstone [31]**:** a metabasite like a greenschist but without foliation.

grossular [29, 113]**:** calcium garnet, formula $Ca_3Al_2(SiO_4)_3$.

groundwater [7]**:** rainwater or seawater that fills the pore space (voids and crevices) in rocks at relatively shallow depths.

H

half-life [136]**:** the time taken for exactly half the amount of a radioactive isotope to decay.

harzburgite [44]**:** peridotite composed of olivine and orthopyroxene. It is a residue of the partial melting of the original peridotite in the Earth's mantle after basaltic magma has been removed.

hematite [15]**:** a common accessory mineral, formula Fe_2O_3. When very fine-grained it imparts a red colour to a rock.

hornblende [32, 114]**:** the most common clinoamphibole with the basic formula $Ca_2(Mg,Fe)_5Si_8O_{22}(OH)_2$ and considerable atomic substitution of Na for Ca, Mg for Ca, Al for Mg, and Al for Si. In addition, K can be present. It commonly forms prismatic black crystals with two directions of cleavage planes intersecting at about 60°.

hornfels [7]**:** a tough, fine-grained, unfoliated metamorphic rock that occurs in contact aureoles.

hydrothermal metamorphism [7]**:** metamorphism caused by hot water as it percolates through permeable rock. The flow of hot water is generally driven by thermal convection induced by heat from nearby magma.

hydroxyl ion [110]**:** a tightly bonded combination of a hydrogen atom and an oxygen atom. It carries a single negative charge and is written $(OH)^-$.

I

igneous rock [1]**:** any rock formed by the cooling and crystallization (hardening) of hot liquid rock called magma. Igneous rocks are also called magmatic rocks.

illite [15]**:** a clay mineral that occurs in shale and slate. It is a very fine-grained, poorly crystallized variety of muscovite.

inclined extinction [123]**:** see parallel extinction.

inclusion (mineral inclusion) [19]**:** a mineral grain enclosed within a larger mineral grain.

index mineral [21]**:** one of the six minerals chlorite, biotite, garnet, staurolite, kyanite and sillimanite that appear in sequence in metapelites as the grade increases in a metamorphic belt.

interference colours [121]**:** the colours (including grey and white) seen with a polarizing microscope in mineral grains in a thin section of rock between crossed polars (XP). They are grouped into so-called orders, named first order, second order etc.

intermediate (igneous rock) [41]**:** a term for an igneous rock whose silica content lies between 52% and 63% by weight, and it includes andesite and diorite.

ion [109]**:** an atom that has either lost one or more electrons, making it positively charged, or gained one or more electrons, making it negatively charged.

ionic chemical bond [109]**:** a strong attraction that binds a positively charged ion to a negatively charged ion inside minerals and other solid materials.

isochron diagram [101]**:** a diagram in which data points for isotope ratios in coexisting minerals fall on a sloping line from which the age of the formation of the minerals can be calculated.

isograd [21]**:** a line on the ground or on a map delineating the first appearance of an index mineral.

isotope [137]**:** a specific kind of atom. It is a unique combination of neutrons and protons whose sum gives the isotopic number. For example, the isotope named potassium-40 (written ^{40}K) contains 19 protons and 21 neutrons (total 40). There are three naturally occurring isotopes of potassium, ^{39}K, ^{40}K and ^{41}K. Each contains 19 protons.

isotopic dating [79, 136]**:** the standard method of measuring the ages of minerals. It is based on the so-called radioactive decay of one isotope (the radioactive parent isotope), at a known rate, to another isotope (the radiogenic daughter isotope), where the amounts of parent and daughter in a mineral can be measured precisely.

isotropic (optical property) [121]**:** a term describing a mineral grain, like garnet, in a thin section of rock that always turns black in XP.

J

jadeite [81, 113]**:** clinopyroxene with the formula $NaAlSi_2O_6$.

K

K-feldspar [115]**:** see potassium feldspar.

kilobar (kbar) [5]**:** A traditional unit of pressure used in studies of metamorphic rocks. 1 kbar is equivalent to 1000 bars (roughly 1000 atmospheres) and corresponds to the pressure at a water depth of 10 km or a rock depth (depending on rock density) of about 3.5 km. The standard (SI) unit of metamorphic pressure is the gigapascal (GPa). 1 GPa = 10 kbar.

Kübler Illite Crystallinity Index [16]**:** an indicator of the temperature reached by shale that has undergone diagenesis or low-grade metamorphism to slate.

kyanite [19, 109]**:** a dense blue silicate mineral with blade-shaped crystals and formula Al_2SiO_5. It is one of three polymorphs of Al_2SiO_5, the others being andalusite and sillimanite.

L

Lava [1]**:** the name for magma when it flows on the land surface or on the seabed.

lawsonite [81]**:** a Ca-rich hydrous silicate mineral formed at high pressure and low temperature.

leucosome [22]**:** the light-coloured component of migmatite.

lherzolite [44]**:** peridotite with orthopyroxene and clinopyroxene.

limestone [1]**:** a sedimentary rock composed of shells and other particles of calcium carbonate ($CaCO_3$) that have become cemented together.

limited atomic substitution [91, 111]**:** atomic substitution where atoms (strictly, ions) of one element can substitute for those of another element only to a limited extent because the two kinds of atom are of different sizes.

lithification [4, 50]**:** the process, usually in the shallow subsurface, by which loose sediment changes into hard sedimentary rock.

142

lithosphere [105]**:** the outer layer of the solid Earth down to a depth of about 100km. Its base is within the mantle and it is defined by a gradual softening of the mantle rock (peridotite) as its temperature rises with depth. At about 100km depth the rock becomes sufficiently hot and weak that it is no longer called lithosphere but asthenosphere. The crust forms an upper layer within the lithosphere.

lithostatic stress [5]**:** see pressure.

M

magma [50]**:** liquid rock, with or without suspended crystals and gas bubbles. On cooling it crystallizes (solidifies) and turns into igneous rock.

magnesite [48]**:** magnesium carbonate, $MgCO_3$.

mantle [1, 103]**:** the rocky part of the Earth's interior lying beneath the crust (i.e. below the Moho). It is thought to be composed largely of the rock peridotite. It continues down to a depth of about 2900km, where the core begins.

marble [3, 24]**:** metamorphosed limestone, recognized by the predominance of calcium carbonate ($CaCO_3$).

marl [29]**:** a sedimentary rock transitional between limestone and shale.

melanosome [22]**:** the dark component of migmatite.

mesoperthite [91]**:** perthite with equal areas of K-feldspar and albite.

meta (as a prefix) [3]**:** a prefix which when combined with the name of a protolith gives the name for the metamorphic rock derived from that protolith, such as metagabbro, metasediment, metashale etc.

metabasite [30]**:** metamorphosed basic igneous rock.

metamorphic aureole [6]**:** see contact aureole.

metamorphic belt (regional metamorphic belt) [6]**:** a large, elongated tract of land where a mountain range once stood but has since been eroded away leaving metamorphic rocks, formerly in the mountain's roots, exposed at the Earth's surface.

metamorphic core complex [79]**:** a generally young, high-grade metamorphic belt thought to have been exhumed through stretching of the lithosphere.

metamorphic facies [39]**:** one of the eight or so 'windows' of pressure and temperature conditions where metamorphism occurs. Most metamorphic facies take the name of the metabasite that forms within them.

metamorphic reaction [54]**:** a chemical reaction involving minerals, and usually also a fluid, accompanying metamorphism.

metamorphic zone [21]**:** an area on the ground or on a map lying between two isograds. For example, the garnet zone occupies the area between the garnet isograd and the staurolite isograd. It corresponds broadly to a temperature interval.

metamorphism [1, 7]**:** the process of change in the sizes, shapes and arrangements of the grains, and usually also in the mineral constituents, of any pre-existing rock, under subsurface conditions beyond the realm of diagenesis.

metapelite [3]**:** metamorphosed shale or mudstone.

metapsammite [3]**:** metamorphosed sandstone.

metasomatism [2, 32]**:** metamorphism in which there is a change to the overall chemical composition of a rock, in addition to possible changes in the amounts of H_2O and CO_2.

metastable [52]**:** in metamorphism it describes the state of a rock at a particular pressure and temperature where the rock has an excess of Gibbs energy. With the input of more energy to provide a metaphorical 'push' the rock can lose the excess energy and change to a stable state.

mica [16]**:** a group name for the common sheet silicates, biotite and muscovite.

mid-ocean ridge [106]**:** see spreading ridge.

migmatite [5, 22]**:** a high-grade gneiss, usually a metapelite, that is a mixture of light-coloured rock composed of quartz and feldspar and dark-coloured rock that is usually biotite-rich. It is thought to be a product of partial melting.

mineral [1]**:** a naturally occurring solid chemical compound. Minerals are the 'ingredients' of rocks; i.e. a rock is an aggregate of one or more minerals. Each mineral has its own distinctive internal crystal structure. The chemical composition of a mineral is fixed or varies within strict limits that are determined by the crystal structure.

mineral assemblage [2]**:** the list of minerals present in a metamorphic or other rock.

mineral lineation [33]**:** a term for a metamorphic rock texture in which prismatic crystals, such as crystals of hornblende, are aligned parallel to each other.

Moho [88, 103]**:** see Mohorovičić discontinuity.

Mohorovičić discontinuity (Moho) [103]**:** the lower boundary of the Earth's crust where the speed of sound (seismic P-waves) increases suddenly from about 6.5km/sec (in the crust) to about 8km/sec in the rocks below (the mantle).

muscovite [15, 115]**:** a sheet silicate with the formula $KAl_2Si_3AlO_{10}(OH)_2$. Along with biotite it is a variety of mica.

mylonite [9]**:** an extremely fine-grained, flinty-looking and thinly banded metamorphic rock produced during dynamic metamorphism in shear zones.

Myr [58]**:** abbreviation for million years.

N

neocrystallization [65]**:** growth of grains of newly formed minerals from a metamorphic reaction, as opposed to the growth of pre-existing minerals, known as recrystallization.

nm [98]**:** nanometre, a unit of length. It is one thousand millionth of a metre (10^{-9} metres).

nucleation (= crystal nucleation) [66]**:** the spontaneous formation of a small crystal that is capable of growing larger.

O

oceanic crust [103]**:** the outer layer of the Earth beneath the deep oceans. Its base is defined by a jump in seismic velocity at the oceanic Moho. It is typically between 5 and 10km thick and it consists largely of the rocks basalt, dolerite and gabbro.

octahedral site [112]**:** in silicate minerals, the space inside an octahedron made by six oxygen atoms in contact. It is commonly filled by an iron or a magnesium atom.

olivine [26, 104, 110]**:** a hard, dense, green glassy-looking mineral with the formula $(Mg,Fe)_2SiO_4$. It forms a solid solution series between the end-members forsterite (Mg_2SiO_4) and fayalite (Fe_2SiO_4).

omphacite [35, 113]**:** green clinopyroxene transitional between augite and jadeite. It is stable only at very high pressures and occurs in the rock eclogite.

ophiolite [104]**:** a large rock unit containing pillow basalt, dykes, gabbro and peridotite that is believed to be former oceanic crust, which has been displaced and now rests on top of, or lies within, continental crust.

order (of interference colour) [121]**:** see interference colour.

orogenesis (orogeny) [3]**:** the process of mountain building (a mountain-building event).

orogenic belt [6]**:** tract of country where orogenesis is now in progress or where it took place in the past. In the second case it is the same thing as a metamorphic belt.

orthoamphibole [114]**:** amphibole with only octahedral sites between the chains, and having the formula $(Mg,Fe)_7Si_8O_{22}(OH)_2$.

orthogneiss [41]**:** gneiss derived from an igneous protolith.

orthopyroxene [33, 113]**:** pyroxene with only octahedral sites between the chains, and having the formula $(Mg,Fe)_2Si_2O_6$, also written $(Mg.Fe)SiO_3$.

P

paragneiss [41]**:** gneiss derived from a sedimentary protolith.

parallel extinction (= straight extinction) [123]**:** the case where the elongate outline, or the trace of a single set of cleavage planes, of a mineral grain in a thin section is orientated perfectly up-and-down or perfectly sideways when the grain is in extinction. Where the orientation at extinction is diagonal (inclined), the mineral is said to have inclined extinction.

partial melting (= anataxis) [ ]**:** a process whereby metamorphism at a high temperature (high grade) turns a rock into a very hot 'slush' of liquid and residual solid grains. If the rock is peridotite, the liquid is usually basaltic, and if the rock is pelitic the liquid is usually granitic.

pelite [14]**:** obsolete term for shale or mudstone, but sometimes used instead of metapelite.

peridotite [1, 44, 104]**:** a rock composed largely of the mineral olivine. It is the main rock of the mantle.

Periodic Table [110]**:** a table in which the chemical elements are listed in the order of their atomic number (the number of protons each element contains), starting with hydrogen. The elements are laid out in rows (periods) and columns (groups) and the position of each element reflects its chemical behaviour.

perthite [91]**:** a variety of feldspar comprising finely interdigitated layers of albite and K-feldspar, formed by the process of exsolution.

petrography [11]**:** the description of rocks. It is the first stage in the overall study of rocks, known as petrology. The second stage, interpretation, is called petrogenesis.

phyllite [3, 16]**:** a foliated metamorphic rock that is transitional between slate and schist. It is usually, though not always, derived from shale during low-grade metamorphism.

pillow basalt [30]**:** basalt that resembles a pile of pillows, each draped on those below. It can only form where basaltic lava is erupted under water. The shapes of the pillows may be preserved during metamorphism.

plagioclase (= plagioclase feldspar) [30, 111]**:** a framework silicate and the most abundant variety of feldspar. It forms a solid solution series between a sodium end-member called albite ($NaAlSi_3O_8$) and a calcium end-member called anorthite ($CaAl_2Si_2O_8$).

plane polarized light [120, 122]**:** see PPL.

planar deformation features (PDFs) [85]**:** a product of damage caused by shock metamorphism, seen as sets of dark parallel bands in quartz and other minerals.

plate (= tectonic plate) [6]**:** one of the dozen or so huge curved 100-km-thick slabs of cool, strong uppermost mantle, known as the lithosphere. Together the plates cover the entire surface of the planet. They move slowly relative to each other. Most plates are partly capped by continental crust, and partly by oceanic crust.

plate tectonics [104]**:** a theory of the Earth based on plates.

platy [65]**:** a term describing crystals with a broad flat shape, like mica.

pleochroism (adjective pleochroic) [19, 123]**:** the case where a mineral grain in thin section, viewed in PPL, changes colour as the microscope stage is turned.

pluton [7]**:** a term commonly used for a body of coarse-grained intrusive igneous rock, such as granite, that is not a sill or dyke, and is usually several kilometres across.

plutonic igneous rock [51]**:** a coarse-grained rock that solidified slowly as a pluton or other large intrusive body.

poikiloblast [19]**:** a porphyroblast with many inclusions.

polarizing microscope [118]**:** a microscope with two polarizing filters and a rotating stage, used to examine thin sections of rock.

polyhedron (adjective polyhedral, plural polyhedra) [62, 112]**:** a three-dimensional shape bounded by a number of flat surfaces. Examples are a tetrahedron (4 surfaces – the minimum), a cube (6), an octahedron (8), a dodecahedron (12) or shapes with any other number of flat surfaces. Polyhedra are the shapes adopted by bubbles in foam, by mineral grains in many metamorphic rocks, and, on the scale of atoms, by clusters of oxygen atoms surrounding positive ions.

polymorph [23, 113]**:** one of two or more different minerals each having the same chemical formula.

porphyroblast [18]**:** a mineral grain in a metamorphic rock, often with a distinct crystal shape, that has grown to a conspicuously larger size than the other grains in the rock.

porphyroclast [42]**:** a mineral grain in mylonite or other cataclastic rock that, having survived crushing, is conspicuously larger than the other grains in the rock.

potential energy [52]**:** the energy contained by an object due to its elevation.

potassium feldspar (K-feldspar) [111]**:** feldspar with the formula $KAlSi_3O_8$. It is an essential mineral in granite.

PPL [11]**:** plane polarized light, produced when only one polarizing filter is in the light path in a polarizing microscope.

prehnite-pumpellyite [39]**:** a pair of hydrous, calcium-rich silicate minerals that form together, for example in basic igneous rocks, at low grade and give their names to a metamorphic facies.

pressure [5]**:** a force that increases with depth and tends to reduce the volume (increase the density) of rock by squashing it equally in all directions. It is also known as lithostatic stress, and is conventionally expressed in units called kilobars (kbar).

prismatic [ ]**:** a term describing crystals (e.g. hornblende, andalusite) with a long, thin shape.

prograde [21, 56]**:** metamorphic changes that develop during increasing grade (temperature) are described as prograde.

protolith [1]**:** the original igneous or sedimentary rock from which a metamorphic rock is derived.

psammite [13]**:** obsolete term for sandstone, sometimes used instead of metapsammite.

pseudomorph [27, 70]**:** a grain of one mineral that has been replaced by another mineral, or minerals, yet has retained its original shape.

pseudotachylite [43, 85]**:** a dark, glassy product of intense dynamic or shock metamorphism.

P-T-t path [77]**:** abbreviation for pressure–temperature–time path, a record of the changing pressure and temperature through time followed by a metamorphic rock during and after orogenesis or subduction.

P-waves [84]**:** primary seismic waves. They are compressional waves (sound waves) transmitted by solid rock and by liquids.

pyrope [82, 113]**:** magnesium garnet, $Mg_3Al_2(SiO_4)_3$.

pyroxenes (= single chain silicates) [33]**:** a group of silicate minerals whose structure is based on infinite parallel chains of linked SiO_4 tetrahedra, giving the repeat unit $(Si_2O_6)^{4-}$. Pyroxene minerals have extensive atomic substitution, and have two directions of cleavage that intersect at about 90°.

pyroxene granulite [113]**:** a kind of metabasite that is dark in colour and composed of the minerals orthopyroxene, clinopyroxene (augite), and Ca-bearing plagioclase.

Q

quartz [109]**:** a common, clear glassy-looking mineral composed of silica, SiO_2, and having a framework silicate structure.

quartz ribbons [11]**:** parallel strips of quartz seen in thin sections of blastomylonite.

quartzite [44]**:** metamorphosed quartz sandstone. Confusingly the same word is used for quartz sandstone that has not been metamorphosed.

R

radioactive [4, 77, 136]**:** an adjective describing an element, or, strictly, an isotope of an element, such as U, Th or K, that is unstable and slowly changing to another element, and releasing heat as it does so.

radiogenic [77]**:** adjective describing a product of the decay of a radioactive isotope. It can refer to the new (daughter) isotope and also the heat released when the decay takes place.

raster pattern [130]**:** the track followed by the focused tip of an electron beam as it scans over a selected rectangular area of a specimen surface in a scanning electron microscope by tracing a series of adjacent parallel lines.

recrystallization [3, 63]**:** see grain growth.

reduction spot [15]**:** a greenish, rounded area on a reddish slate where the red pigment, hematite, has locally been removed by chemical reduction of iron.

refractive index [119]**:** an optical property of a transparent material. It is the ratio of the speed of light in a vacuum to the speed of light through that material.

regional metamorphism [3]**:** metamorphism seen on a regional scale. It is usually a result of processes within the roots of actively forming mountain ranges.

relict igneous texture [40]**:** a distribution of mineral grains in a metamorphic rock that has been inherited from the texture of an igneous protolith.

relief [119]**:** a term denoting whether a mineral grain in a thin section of rock in PPL appears rough (high relief) or flat and smooth (low relief).

restite [22]**:** the unmelted grains following partial melting and removal of the liquid.

retrograde metamorphism [27, 57]**:** metamorphism that takes place while the temperature is falling.

rhombic dodecahedron [18]**:** a crystal shape with twelve diamond-shaped faces commonly displayed by garnet.

rhyolite [41]**:** volcanic rock produced when granitic magma erupts as lava.

ringwoodite [87]**:** a dense, blue-coloured high-pressure polymorph of $(Mg,Fe)_2SiO_4$ formed by shock metamorphism of olivine.

rock cycle [1, 50]**:** a flow chart showing the processes (weathering, erosion, sediment transport, sediment deposition, lithification, metamorphism, partial melting) that lead to the formation of one kind of rock from another, on the surface and within the continental crust.

rutile [36]**:** an accessory mineral, formula TiO_2, common in eclogite.

S

sandstone [1]**:** a sedimentary rock made from sand grains. It is usually dominated by grains of quartz, making its overall chemical composition silica rich (SiO_2-rich).

sapphire [72]**:** see corundum.

scanning electron microscope (SEM) [15, 119, 130]**:** an instrument that utilizes a moving electron beam to create magnified images, and chemical maps, of a specimen's surface.

schist [3]**:** a foliated metamorphic rock that has a wavy sheen and can be split into flat or gently curved pieces. It usually, but not always, contains abundant mica and is derived from shale during medium-grade metamorphism.

schistosity [17]**:** the distinctive characteristic of schist whereby the rock tends to break into flat or gently curved pieces due to the alignment of the mineral grains, usually mica.

sedimentary rock [1]**:** any rock formed by the accumulation and subsequent lithification (binding) of fragments of material, or formed as a chemical precipitate. It can form on the surface, or on the seabed or a lake bed.

semipelite [22]**:** a metamorphic rock transitional between metapelite and metapsammite, derived from a protolith of muddy sandstone or siltstone.

serpentine [27, 115]**:** a green, waxy-looking sheet silicate with a formula corresponding to a combination of talc and magnesium hydroxide, $Mg_3Si_4O_{10}(OH)_2.3(Mg(OH)_2)$.

serpentinite [46]**:** a metamorphic rock composed largely of the mineral serpentine. Its protolith is peridotite.

shale [1, 13]**:** a sedimentary rock composed of lithified mud that is fissile, i.e. that can be split into thin wafers parallel to the original sedimentary layering. The term is used loosely in this book to include mudstone, which is not fissile.

shatter cones [84]**:** sets of aligned, cone-shaped fractures caused by shock metamorphism.

shear zone [9]**:** a fault zone at considerable depth where sheet-like bodies of intensely flattened and deformed rock, such as mylonite, are produced.

sheet silicates [114]**:** a group of silicate minerals whose structure is based on flat sheets of linked SiO_4 tetrahedra, giving the repeat unit $(Si_4O_{10}(OH)_2)^{6-}$. Sheet silicates include talc, the micas, chlorite and serpentine.

shock metamorphism [9]: metamorphism of a swift and violent nature caused by the passage of intense shock waves from the impact of a giant meteorite.

sieve texture [95]: a texture in which a high proportion of inclusions of one mineral occur inside another mineral, such that the inclusions resemble the holes in a sieve.

silicate minerals [112]: minerals whose chemistry is dominated by silicon and oxygen.

sill [30]: a sheet-like intrusion of igneous rock that is parallel to layering in the country rock.

sillimanite [19, 113]: a silicate with spiky prismatic crystals and formula Al_2SiO_5. It is one of three polymorphs of Al_2SiO_5, the others being andalusite and kyanite.

single chain silicate [113]: see pyroxene.

slate [3, 14]: a very fine-grained foliated metamorphic rock that displays slaty cleavage. Most slate is derived from shale during very low-grade metamorphism.

slate belt [14, 15]: all or part of a metamorphic belt where slate dominates.

slaty cleavage [15]: the potential for slate to be split into thin flat sheets. This is caused by an abundance of microscopic plates of mica that are orientated parallel to one another.

soapstone [47]: metamorphosed peridotite composed largely of talc.

solid solution series [110]: an alternative term for atomic substitution. The full range of compositions of a mineral exhibiting solid solution is called a solid solution series. The simple mineral formulae at the limits of a solid solution series are called end-members.

solidus [107]: the temperature at which a mineral or rock begins to melt as it gets hotter. Its value depends on pressure.

solvus [91]: an arch-shaped curve on a diagram showing the extent of limited atomic substitution between two end-members over a range of temperatures.

spherule bed [86]: a sedimentary layer composed of small spherical tektites called spherules.

spilite [31]: greenstone resulting from the hydrothermal alteration of basalt or dolerite.

spreading ridge (mid-ocean ridge) [8, 106]: a submarine ridge of mountains or shallow seabed where two plates are moving apart and new lithosphere, capped with oceanic crust, is being created.

stability [52]: see equilibrium.

staurolite [19, 113]: a dense brown silicate with blocky, sometimes cross-shaped crystals and a formula approximating 2 units of Al_2SiO_5 with (Fe,Mg)O(OH).

straight extinction [123]: see parallel extinction.

strain [5]: a change in shape. A body of rock becomes strained when subjected to a sufficient amount of directed stress.

strain shadowing [36, 68, 126]: a feature seen in thin sections of mineral grains that have been strained. As the microscope stage is rotated in XP a shadow appears to pass over a grain as misorientated subgrains within it pass, in turn, through extinction. It is common in quartz, where it is also called undulose extinction.

stretching lineation [42]: a metamorphic texture where grains have been smeared out into long parallel, and often flattened, streaks.

subduction zone [107]: a boundary between converging plates, where one plate, which is usually capped with oceanic crust, is subducted (slides back into the mantle) beneath the other plate.

suevite [84]: breccia with shock-melted fragments found in and around impact craters.

supersaturated (fluid) [66]: a solution having excess Gibbs energy because the concentration of dissolved material at a particular pressure and temperature is higher than it would be if it were in equilibrium.

symplectite [95]: a fine-scale intergrowth of two or more minerals, for example plagioclase and clinopyroxene, usually formed during retrograde metamorphism or in response to decompression.

S-waves [107]: secondary seismic waves. They are transmitted by sideways shaking motions and cannot pass through liquids. They travel more slowly than P-waves.

T

talc [45, 115]: a sheet silicate with the formula $Mg_3Si_4O_{10}(OH)_2$.

tartan twinning [4, 105]: a distinctive criss-cross pattern seen sometimes in K-feldspar in thin sections under XP.

tectonic plate: see plate.

tektite [86]: a bead of frozen, shock-melted liquid launched from an impact site.

tetrahedral site [112]: in silicate minerals, the space inside a tetrahedron made from four touching oxygen atoms. It is occupied by an atom of Si or Al.

tetrahedron (plural tetrahedra) [112]: a three-dimensional shape bounded by four flat, triangular-shaped surfaces.

texture [1]: the sizes, the shapes, and the arrangement of mineral grains in a rock. The arrangement of grains includes whether or not they have a preferred orientation, and whether or not the grains of each mineral are uniformly distributed.

thermal aureole (thermal metamorphism) [7]: see contact aureole (contact metamorphism).

thin section [11, 118]: a slice of rock 30 microns thick glued onto a glass microscope slide, specially prepared for examining the rock through a polarizing microscope.

titanite [33]: an accessory mineral, formula $CaTiSiO_5$.

triangular diagram [29]: an equilateral triangle showing graphically the composition of a rock or mineral in terms of three components, one at each apex.

tremolite [27, 114]: the magnesium end-member of clinoamphibole, $Ca_2Mg_5Si_8O_{22}(OH)_2$.

tourmaline [117]: a coloured accessory silicate mineral containing the element boron

twinkling [48]: the noticeable change in relief of a mineral grain in thin section as the microscope stage is turned.

twin lamellae (in plagioclase) [124]: parallel grey stripes caused by twinning seen in grains of plagioclase in a thin section in XP. Twin lamellae of a different kind may be seen in calcite and dolomite.

twinned crystal [19]: a combination of two or more separate crystals that have grown together to make a single object in a way determined by the pattern of the internal crystal structure.

U

ultrabasic [44]: a term for an igneous rock whose silica content is less than 45% by weight.

ultra-high-pressure metamorphism [82]: metamorphism under conditions where coesite is stable.

undulose extinction [68, 126]: see strain shadowing.

V

vein [47]: a fracture plane in a rock that has gaped open as a result of pull-apart forces, and has subsequently been filled with a mineral. The mineral is related to the host rock, so, for example, calcite veins are normal in limestone or marble, quartz veins are common in sandstone or psammite, and white asbestos veins occur in serpentinite.

Virtual Microscope [118]: a website (virtualmicroscope.org) with a 'library' of rock thin sections to examine in PPL and XP.

volcanic arc [107]: a line, usually gently curved, of volcanoes that have erupted above a subduction zone.

W

water-deficient (rock) [56]: a rock which, under a particular range of pressure and temperature conditions, would spontaneously react with water if water were present.

weathering [50]: the processes of mechanical disaggregation (physical weathering) and chemical alteration, e.g. to clay minerals (chemical weathering), of a rock exposed to the weather.

wollastonite [70]: a white calcium silicate, formula $CaSiO_3$.

X

xenolith [40]: a piece of any rock enclosed by an igneous rock, usually a detached fragment of country rock engulfed by magma before it solidified.

XP [11, 120]: abbreviation for crossed polars, which is where both polarizing filters in the microscope are in the light path.

X-ray powder diffractometry (XRD) [15]: a technique that uses X-rays for identifying minerals alone or in a rock. It involves first grinding the sample to powder.

Z

zeolite [39]: a group of hydrous framework silicates that grow under very low-grade conditions, for example in basic igneous rocks. Zeolite lends its name to a metamorphic facies.

zone [5]: see metamorphic zone.

Further reading

The following publications develop the subject of metamorphism beyond the level of this book:

Best, Myron G. (2013) *Igneous and Metamorphic Petrology.*Oxford: Wiley-Blackwell, Second edition, 752pp.

Fettes, D. and Desmons, J. (eds) (2011) *Metamorphic Rocks: A Classification and Glossary of Terms.* Cambridge: Cambridge University Press, 244pp.

Mason, Roger (1990) *Petrology of the Metamorphic Rocks.* London: Unwin Hyman Ltd, Second edition, 240pp.

Vernon, R.H. and Clarke, G.L. (2008) *Principles of Metamorphic Petrology.* Cambridge: Cambridge University Press, 446pp.

Yardley, B.W.D. (1989) *An introduction to metamorphic petrology.* Harlow, UK: Longman, 248pp.

Yardley, B.W.D., MacKenzie, W.S. and Guilford, C. (1990) *Atlas of metamorphic rocks and their textures.* Harlow, UK: Longman, 120pp.

In addition to these books, many websites provide excellent images of metamorphic rocks and accounts of how they are classified and how they were made. Among these websites are Wikipedia.org and sandatlas.org.